Innovation, Technology, and Knowledge Management

Series Editor
Elias G. Carayannis, George Washington University
Washington, DC, USA

This series highlights emerging research and practice at the dynamic intersection of innovation, technology, and knowledge management, where individuals, organizations, industries, regions, and nations are harnessing creativity and invention to achieve and sustain growth. Volumes in the series explore the impact of innovation at the "macro" (economies, markets), "meso" (industries, firms), and "micro" levels (teams, individuals), drawing from such related disciplines as finance, organizational psychology, R&D, science policy, information systems, and strategy, with the underlying theme that in order for innovation to be useful it must involve the sharing and application of knowledge.

More information about this series at http://www.springer.com/series/8124

Suchismita Satapathy • Debesh Mishra
Arturo Realyvásquez Vargas

Innovation in Agriculture with IoT and AI

Springer

Suchismita Satapathy
Mechanical Engineering
KIIT University
Bhubaneswar, Odisha, India

Debesh Mishra
School of Mechanical Engineering
KIIT University
Bhubaneswar, Odisha, India

Arturo Realyvásquez Vargas
Departamento de Ingeniería Industria
Instituto Tecnologico de Tijuana
Tijuana, Mexico

ISSN 2197-5698 ISSN 2197-5701 (electronic)
Innovation, Technology, and Knowledge Management
ISBN 978-3-030-88827-5 ISBN 978-3-030-88828-2 (eBook)
https://doi.org/10.1007/978-3-030-88828-2

This Springer imprint is published by the registered company Springer Nature Switzerland AG
The registered company address is: Gewerbestrasse 11, 6330 Cham, Switzerland

Preface

This book has been organized for easy readability and portability. It includes seven chapters. Chapter 1 provides a brief report about the prevalent problems in the global agri-sectors. Chapter 2 gives a literature review on global agri-sector. Chapter 3 explores the features of Internet of Things (IoT) to resolve agricultural problems. Chapter 4 investigates measurement and prediction of mental stress through innovative equipment in agriculture, and Chap. 5 gives an insight into personal protective equipment for farmers. Chapter 6 explains food safety and IoT-based solution. Finally, Chap. 7 ends with future findings. Each chapter is well organized and contains case studies. It is an excellent guide for innovations in agri-sector with information technology, which helps both academicians and researchers in their research.

Agri-sector problems are multifaced and multidisciplinary in nature. They occur globally and are important topics for discussion among scientist, stake holders, and researchers. Food safety, agrarian safety, sustainability in farming, health and safety issues of farmers, crop protection, and developments in agricultural equipment and machinery have been attracting worldwide focus to promote agri-business as it also deals with the socio-economic and environmental hazards. With the use of sensitive IoT-based devices and artificial intelligence (AI) techniques, innovative equipment and machines can help resolve agri-sector problems. This authored book, *Innovation in Agriculture with IoT and AI,* contains seven chapters that are distinguished but interrelated to each other. Each of these extended chapters has undergone full double-blind peer review before being selected for this book. These very informative chapters will definitely be helpful to researchers, academicians, and industry persons as well.

Agriculture has been considered an unsafe industry, by seeing the past cases of accidents, physical hazards, and mental health conditions of farmers. Upsetting life occasions have been significant in fostering mental issues among ranchers. Majority of the non-industrial nations the self-destructive instances of agrarian specialists have been expanding inferable from number of baffling of disappointing elements, upsetting their typical psychological well-being conditions. Subsequently, distinguishing and focusing on such danger elements might help in proper intercession methodologies. Emotional well-being issues among ranchers occur when their reactions to the horrible response from unreasonable pressing factors or different requests set on them. In addition, horticulture is worried about an assortment of decisions just as vulnerabilities. Attributable to the variety of climate from one season to another, vacillation on cultivating materials costs, soil debasements, unviable yields, suffocation of harvests by weeds, bugs harming the harvests, and changing of environments is a vulnerability the ranchers must overcome. As the researchers are unable to innovate the new equipment to resolve the problems in agriculture only solved the problems with the help of computational methods.

Yet, development and cutting-edge innovations for all intents and purposes have settled horticultural issues by alleviating physical and mental responsibility of agri-farmers with IoT and AI techniques. IoT and AI provide simple approaches to tackle any difficult task without hard labour. Such technologies can additionally be used to keep ranchers safe and assist them with. Safety and shielding supplies have settled numerous issues such as cultivating, ploughing, and plucking vegetables. Likewise, IoT and AI assist farmers in guarding their fields and food items from wild animals. For occupational well-being and security in the agri-sector, AI and IoT provide imaginative defensive guides. This is another area for inventive research and new types of gear for cultivating.

Bhubaneswar, Odisha, India — Suchismita Satapathy
Bhubaneswar, Odisha, India — Debesh Mishra
Tijuana, Mexico — Arturo Realyvásquez Vargas

Contents

About the Authors

Suchismita Satapathy PhD, is an associate professor in the School of Mechanical Sciences, KIIT University, Bhubaneswar, India. She has published more than 130 articles in national and international journals and conferences. She has also published many books and e-books for academic and research purposes. Her areas of interest include production operation management, operation research, acoustics, sustainability, and supply chain management. With more than 15 years of teaching and research experience, she has guided many PhD (2 completed, 4 ongoing), MTech (20), and BTech (49) students. Suchismita has published books such as *Production operation management* (Stadium Press) and *MCDM methods for waste management* with CRC Press, and many of her projects are ongoing. ORCID:https://orcid.org/0000-0002-4805-1793,Scopus Author ID: 55085975400

Debesh Mishra PhD, has more than 10 years of teaching experience at different engineering colleges at both undergraduate and diploma levels. In addition, he has published more than 20 articles in different reputed journals and international conferences. Also, 10 book chapters are credited to him. His areas of interest include ergonomics, occupational health and safety, sustainability, supply chain management, and optimization techniques. Dr Mishra obtained his PhD in mechanical engineering from KIIT Deemed to be University (Odisha), India. He completed MTech with specialization in industrial engineering and management, and BE in mechanical engineering. 57204262910, https://orcid.org/0000-0003-2665-1365

Arturo Realyvásquez Vargas is a full-time professor in the Department of Industrial Engineering at Tecnológico Nacional de México/Instituto Tecnológico de Tijuana in Mexico. He received his master's degree in industrial engineering and PhD in engineering sciences from the Autonomous University of Ciudad Juárez in Mexico, and also a PhD in innovation in product engineering and industrial process from the University of La Rioja (Spain). His main research areas are optimization of industrial processes, lean manufacturing, and ergonomics. Also, he is an active member of the Society of Ergonomists of Mexico Civil Association (Sociedad de Ergonomistas de México, SEMAC) and the Network of Optimization of Industrial

Processes (Red de Optimización de Procesos Industriales, ROPRIN). Currently, Dr Realyvásquez is a National Researcher recognized by the National Council of Science & Technology of Mexico (CONACYT) as Candidate. Dr Realyvásquez is author/co-author of approximately 12 papers published in indexed journals like the *Journal Citation Reports*. He has attended international conferences and congresses in Mexico as well as in the United States. Dr Realyvásquez has supervised more than 20 bachelor theses and 5 master's theses. Dr Realyvásquez is author of one book published by the international publisher Springer, which is related to ergonomics. Also, Dr Realyvásquez has edited two books for IGI Global, all of them related to ergonomics. ORCID: https://orcid.org/0000-0003-2825-2595. Scopus Author ID: 56167726800.

Abbreviations

FL	Fuzzy Logic
SWOT	Strengths, Weaknesses, Opportunities and Threats
PADETES	Participatory Demonstration and Training Extension System
PES	Participatory Extension System
NALEP	National Agricultural Livestock targeting Extension Programme
FFS	Farmer Field Schools
MCDM	Multi-Criteria Decision-Making
PDA	Positive-Distance from Average
NDA	Negative-Distance from Average
MT/Year	Million-Tonnes/Year
SI_1	Social-Impacts
ECI_2	Economical-Impacts
$ENVI_3$	Environmental-Impacts
TFNs	Trapezoidal Fuzzy-Numbers
WMA	Waste-Management Alternatives
ANFIS	Adaptive Neural-Fuzzy Inference System
ANN	Artificial Neural-Network
FAO	Food and Agriculture Organization
AI	Artificial Intelligence
SWARA	Stepwise Weight Assessment Ratio Analysis
IoT	Internet of Things
OHS	Occupational Health and Safety
MSDs	Musculoskeletal Disorders
WRMSDs	Work-Related Musculoskeletal Disorders
QEC	Quick Exposure Check
RULA	Rapid Upper Limb Assessment
REBA	Rapid Entire Body Assessment
OWAS	Ovako Working-Posture Analysing-System
MIMO	Multi Input and Output
WSNs	Wireless Sensor Networks
AA	Agent Architectures
EC	Evolutionary Computing

List of Figures

List of Tables

Chapter 1
The Prevailing Problems in the Global Agri-Sectors

1.1 Introduction

In the recent-years, the agri-sectors all over the world have undergone with considerable transformations and changes the day-to-day work-activities. Further, there has been a common perception of large-farms to be more prosperous in adapting as compared to the family-run small-farms. Moreover, intensification of farm-related practices has occurred with regard to the reduced profits and returns in certain farming, and has diminished the controllable sense in exercising the farm-work and has leaded into diversification of farm-activities. However, being the growing industrial-regulation, the agri-sector's mechanization has significantly influenced in the qualitative transformation of day-to-day work-activities of farmers and the whole farming community around the globe.

Although through farming adequate incomes can be generated, but the biggest challenge that occurs is the short fall of farm labour in most of the cases. Owing to the expected returns in farming, irrespective of the hard work put forth by the farmers, they usually pass through a number of factors that influence their mental well-being to choose farming as their primary occupation and to rely on it. Most of the farmers around the globe look for other kind of tasks that are not related to agriculture in order to run their families. The farmers have combined working on their own-farms with seasonal contract-work, and informal-systems of labour-exchange. Workloads of other farms have been intensified with others been forced-out of business. Now a day's majority of young individuals are deterred from agriculture without following their farm-based parents, which has increased significant pressures upon the remaining farming-communities. The developments of stresses in farm-based work have a number of effects upon both physical as well as mental well-being of the farm workers that include lack of sleep, back-problems, work-related worries, irritability and down-feeling. The farmers normally get worried about the finances and future when they suffer of some occupational-related illness with

S. Satapathy et al., *Innovation in Agriculture with IoT and AI*,
Innovation, Technology, and Knowledge Management,
https://doi.org/10.1007/978-3-030-88828-2_1

considerable higher workload intensities and non-controllable aspects in their occupation (such as seasonality and disease).

Although the farm-based stresses are not always framed in terms of physical or mental health-conditions, but sometimes considered as a constant or positive force that motivates the individuals to find solutions to such issues. Apart from intrinsic aspects of farm-related stresses, such as seasonality and adverse-weather conditions' difficulties, the extrinsic aspects of agriculture-based stresses have been a major issue for the farmers taking more strategic as well as management functions. These stressors include the existing political and legislative framework, perceptions on farmers, and the competitive prices that could be obtained for agri-based/agricultural products. For different range of roles in agriculture, the work-related aspects of stresses have been emerging as important concern that are known to be developing owing to the potentially unsafe farming-practices, work-related pressures and loads, the work-activities to particular type of farming, and the associated tensions with farms, family and others. However, the agriculture-based work-related aspects involve a combination of effects on both physical as well as mental health-conditions, which consequently can have measurable impact on farming including exhaustion in work and injuries in workplaces. Therefore, in order to avoid the negative consequences of agriculture-based problems such as work-related loads and stresses, it becomes important worldwide to address such incurring dangerous as well as hazardous situations that can be eliminated with sufficient and proper mechanization techniques for the wide farming-community for a prosperous and better agriculture.

1.2 Agri-Sectors Problems All Over the World

In most of the developing countries, the parents are usually forced to sell their agricultural lands due to poverty and the young generation often find difficulties to assess the lands owing to increasing degradation of lands (FAO, 2011a), even it becomes a greater challenge in obtaining necessary capital for purchasing of lands (FAO, 2011b). Over the next decades, the agricultural production is likely to get affected through environmental changes with an undefined health-outcomes, including agriculture-based yields and crop-quality (Tuomisto et al., 2017). The increasing continuation in world population poses more challenges to our food-systems and well-being, such as climate-changes, magnified ozone-levels, alteration in water-availability, carbon-dioxide fertilization, soil-degradation, deforestation and land-use that can substantially influence the agricultural production-levels. In the absent of widespread as well as implementation of adaptation in addition to the mitigation-strategies for overcoming of the consequences of environmental changes, the global food-security, health as well as well-being could be affected significantly (IPCC, 2014).

The socioeconomic status has been revealed to be influencing person's intake of fruits and vegetables with higher income countries having greater consumption-rate compared to the countries with lower incomes (Miller et al., 2016) and also, with wealthier countries of more consumption than the poor neighborhoods (Dubowitz et al., 2008; Pessoa et al., 2015). The crops are sensitive to changes in environment-conditions (Backlund et al., 2008), and these environment-based changes also have considerable impacts on agricultural production (Porter et al., 2014; Smith et al., 2014). The crop-yields are influenced directly through alteration in temperature and water-availability (Lobell & Gourdji, 2012). As faster crop-growth take place due to higher temperature, this in-turn lowers the yields by reduction in cropping-seasons. The changes in climatic-conditions also increase the crop-losses in addition to damages due to fungi, pests, weeds, and pathogens (Bebber et al., 2013; Flood, 2010). Other than the direct-effects, under hot climatic-conditions with considerable augmented temperatures may also affect in the yielding of vegetables, fruits and other agri-products indirectly due to reduction in labour-productivity of farmers that in turn affect the overall agriculture-based productivity. The climatic variability along with other external factors may have either direct or indirect impact on the agriculture (Fig. 1.1). For instance, the consequences of rising temperature may increase the ozone-formation at ground-level or troposphere-level with considerable reduction in photosynthesis and plant-growth (Ainsworth & Ort, 2010).

For the areas where irrigation water is provided by aquifers, the water-resources' depletion can have negative agriculture-based production's impact. The aquifers' depletion is linked to changes in precipitation-levels, river's exhaustion and augmented water-use, and the precipitation being more in higher latitudes as compared to the regions with 'lower mid-latitude' (Bates et al., 2008). The rural farmers with low economic backgrounds in the arid as well as semi-arid tropics including the "African and Asian mega-deltas" are most vulnerable to water-availability changes. Particularly in the producing countries, the decline of aquifers is contributed by the international food-trade (Dalin et al., 2017). The land has been lost due to soil-erosion and soil-pollution causing a serious threat to global agriculture (Cameron et al., 2015; Gibbs & Salmon, 2015). Other factors that have great impact on the land includes "soil's acidification due to acid rains or by using synthetic

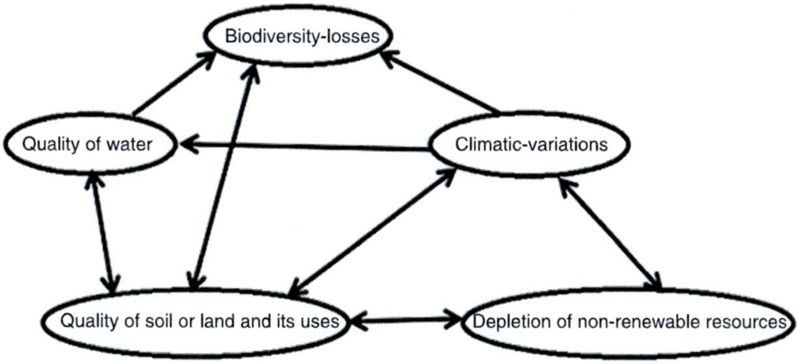

Fig. 1.1 Linkages between changes in external agricultural factors

nitrogen-fertilizers (Klimont et al., 2013); and presence of pollinators (Goulson et al., 2015; Smith et al., 2015). The enhancement in the resilience of 'agro-ecosystems' to environment-related changes can be achieved by maintaining a high biodiversity level (Koohafkan et al., 2012; Lin, 2011), and the farms with a higher biodiversity levels in low-income settings have been found to be more resilient to climate-disasters, such as droughts and hurricanes (Altieri et al., 2015; Meldrum et al., 2013). Erisman et al. (2008) have revealed of agricultural emissions in "Western Europe" that accounted for "approximately 30% of all acidifying-emissions, and 90% of ammonia-emissions". The losses in biodiversity are caused greatly due to industrialization of agriculture, resulting in simplified agro-ecosystems, decreased crops' number and crop-varieties' grown, augmented utilization of chemical fertilizers and pesticides, and conversion of natural-forests to agricultural lands. The enhanced agricultural product's demand is putting pressures in tropical regions to convert forests into agricultural land (Laurance et al., 2014). However, higher on-farm biodiversity has been reported in extensive farming-systems, such as "organic farming-systems" in comparison to intensive farming-systems (Bengtsson et al., 2005; Tuomisto et al., 2012).

The farmers as well as societies have many possibilities for adapting and mitigating of issues related to changes in environment and others (FAO, 2010; FAO, 2012), which can happen at different levels and ranges, including alteration of farm-management practices, such as change of crop-varieties, planting-times, irrigation-practices and residue-management, or agricultural relocation to new-areas (Challinor et al., 2014). Other practices include: mitigation of 'green-house-gases' emission' (Altieri & Nicholls, 2017); novel plant-breeding technologies (Hu & Xiong, 2014); having higher-concentrations and more bio-availability of micro-nutrients (Bhullar & Gruissem, 2013); use of precision farm-technologies like "Small unmanned aerial-systems" (Zhang & Kovacs, 2012); robot's utilization in agriculture (Bogue & Bogue, 2016); and the use of artificial-lighting in case of indoor-farming systems (Cheng, 2014); respectively. During the first half of the twentieth-century, the agriculture-based issues were primarily related to crop-sequence diversity, soil's organic-matter management and management of biological-processes. However, in the "1940s and 1950s", a different view played important role: use of factory-made fertilizers and pesticides for high yields (Drinkwater & Snapp, 2007; Savary, 2014). The industrial agriculture increased significant agriculture-based productivity and also, helped in addressing the food-security issues by considerable transformation limitation of natural-ecosystem into agriculture (Green et al., 2005; Tilman et al., 2001). The new model of agriculture includes the availability and usage of cheaper fossil-fuels, manufactured-inputs based on fossil-fuels, irrigation, mechanization and use of improved-cultivars (Gomiero et al., 2011; Pretty, 2008). The utilization of inputs can be helpful in boosting yield-per-hectare, while the labour-productivity can be enhanced through infrastructure-development (for example: irrigation) and machinery (Arizpe et al., 2011). However, over the past few decades, the agricultural 'industrialization and intensification' have been recognized in causing of significant destructive environment-based impacts.

1.3 Agri-Sectors Problems Due to Physical-Stress

For repetitive lifting-tasks taking place for an extended time-period, a sustained-strength with endurance is required, and the respond of body is observed to be enhancement in heart-rate followed by more amount of oxygen's breathing (Randall & Jeter, 2003). By exceeding the physiological-limits, the occurrence of fatigue and subsequent stress may occur in the cardiovascular-system of some persons, leading to heart-failure, in severe cases (Astrand & Rodahl, 2003). Different related-parameters of mental-stress that influence various physiological-functions included: "blood-pressure, heart-rate, and catecholamine as well as cortisol secretion". However, determining the possible-linkage between psychosocial-stress and several physical health-outcomes has been a subject of interest (Krantz et al., 2004). Usually, repetitive as well as intensive physical-efforts are associated with agricultural activities (Vyas, 2012), which may cause physical-discomfort directly related to the muscular-loads (Vink & Hallbeck, 2012). Moreover, the weeding activities in agriculture have been revealed to cause back-pain among female-workers (Vyas, 2012). The study made by De Castro et al. (2014) revealed that the farmers reported of suffering from physical-discomfort as a function of their performed lighter as well as heavier work-activities. Ma Elena (2009) has revealed of the rural-women in contributing a substantial-share in the agricultural sectors of Philippines, and were subjected to adverse health-consequences including female-reproductive health-risks because of the uses and exposures to unsafe agro-chemicals, agriculture-related accidents, physical-injuries, ergonomic-problems owing to the utilization of men-friendly tools and technology, and nutrition-deficiencies resulting from poverty and over-work.

Moreover, a series of symptoms associated with distress and interfering the personal-functions represent the mental-disorders (WHO, 1992), which include different diseases, including depressions, bipolar-affective disorders, anxiety-disorders, schizophrenia, stresses, substance-abuse related disorders, sleep-wake cycle-related disorders, mental-deficiencies and disorders hindering developments (WHO, 2013a). These disorders were reported to be related with work-function, work-activities type and the work-conditions (WHO, 2013b). The rural-settings have been revealed to be more in the cases of mental ill-health as compared to the urban-settings (Centre for Applied Psychological Research, 2012; Jenkins et al., 2012; Kilkkinen et al., 2007; Lipscomb et al., 2007). A higher prevalence of mental ill-health was reported among rural-based female agricultural-workers in different earlier studies, such as in Ethiopia (Hanlon et al., 2010), Australia (Fisher et al., 2010), India (Shidhaye & Patel, 2010), and Taiwan (Tsao et al., 2015), respectively.

1.4 Agri-Sectors Problems Due to Mental Stress

The issues related to mental health or mental stress among farmers have been a serious and identified health concerns (Hagen et al., 2019), and the farm workers are more commonly influenced with "psychological health-disturbances" (Yazd et al., 2019). The individual health of farmers can be influenced negatively with their ill mental health (Carvajal et al., 2014; Demos et al., 2013; Linn & Husaini, 1987), their home-life (Chaney & Torres, 2017; Olson & Schellenberg, 1986), farm's productivity-levels (Gorgievski-Duijvesteijn et al., 2000), and health and welfare of farm-based animals (Andrade et al., 2014). As the agriculture-based industry is contributing to the global-economy by one-third of individuals (McIntyre et al., 2009), so the existence of poor mental health could have significant negative impact globally on economical production and productivity, and human as well as animal health. Hence, for the global health-concern, ensuring of the mental health of farm workers has become very essential.

Figure 1.2 illustrates the stressful work-activities of farmers that may have negative impacts on mental health. Based on an estimation, every single person among four persons in the world suffer with the issue of mental health every year (WHO, 2001). Also, the farmers' suffering of mental illness as compared to the general public was reported to be higher (Fraser et al., 2005). Pichorner et al. (2017) studied

Fig. 1.2 Stressful work-activities of farmers

the suckler cow-farming in the mountain-areas of Carinthia (Austria), and found the most labor-intensive activities as manure-removals with animal-grouping, feeding and food-preparation. However, mechanization reduced the requirements of working-time as well as the physical-stresses. In a study made by Wilhelmsson et al. (2021) on "Swedish pig transport-drivers", the high arm-positions and more frequencies of shoulder-problems were revealed to put excessive shoulder's physical-load.

1.5 Food Safety

The worldwide agricultural sectors in some rural areas face new challenges to find market for effective selling of their products. These agricultural marketing issues have been a serious concern throughout the world. The increasing agricultural production development has brought its significance in the global agri-sectors. So far agribusiness is concerned; it includes collective business-activities covering the agricultural inputs' supply, production as well as transformation of agri-products and their distribution to the end-users. It has been one of the key generators of employment in addition to income throughout the world. In addition to inadequate education, training, and unsuccessful development in local technologies, the confluence of a variety of local and global drivers has also led to progressing and unpredictable transforms in the functions and structures of worldwide agri-food markets. The agriculture sectors in some rural areas face new challenges with regard to the safety of their products. These agricultural food safety issues have been a serious concern throughout the world.

Moreover, the challenges and opportunities offered in the liberalized trade-regime by the global-markets are also need to be addressed for enabling the farming-community to get benefited from the new global-market. Therefore, the transition to sustainable food and agricultural systems has become essential to manage effectively the global agri-food markets in order to support the expected population-growth and to ensure in universal-access to adequate, safe as well as nutritional food.

1.6 Research Questions

In view of the agricultural pressures' growth in different countries (such as declination in farm-productivity, deterioration in qualities of water and soil, adverse impacts of bad-weather, and other work-related pressures), the understanding of risk-factors that can have negative effect on the farmers' mental health has been getting more importance in order to improve the prevention-effort's efficiency. Hence, it becomes essential in understanding the potential key-risks affecting the smoother growth of overall agriculture as well as the farming-community around the globe. Moreover, throughout the world, more priority-based research needs to be done with regard to the mental health issues with the farmers and their community-groups, which has been primarily recommended by mental health related experts, and other governmental-agencies (McIntyre et al., 2009). Although there exist a multitude of different studies across the world that have examined the incurred stresses in farm-based populations, there are no methodological-guidelines to measure the individual level outcomes with regard to these stresses, which could help the emerging researchers in findings and building of more advanced technologies to reduce the physical as well as mental ill-health conditions of farmers more efficiently.

The objectives of this work were in: (a) providing a descriptive analysis of the pertaining agricultural work-related stresses outcome in the worldwide farming-community; (b) describing the possible international scope in the technological aspects; and (c) highlighting the intervention-strategies for effective and efficient evaluation of workloads and mental stress in agriculture in order to help in the identification of achieved success and the implementing limitations for the global farming-community.

Therefore, in order to achieve the objectives of this work, the following research questions were formulated.

- What are the existing work-strategies in the worldwide agriculture?
- How the workload in agri-sectors affect the farming-community?
- Do the climatic-conditions affect the farm-activities and the farming-community, and if so, then how and to what extent?
- What are the consequences of work burdens on the mental health conditions of farmers?
- What safety practices have been considered in the global agri-sectors with regard to agricultural foods?
- What are the existing techniques and technologies to resolve the ongoing agricultural problems, and what may be other technological options?

1.7 Important Points

There occurs a complex relationship between the environmental, social, and economic factors that affect the worldwide agri-sectors, and the work-related stresses in farming greatly influence the farmers' physical as well as mental well-being. With regard to some ongoing problems in the agri-sectors, such as pesticide-exposure, finance-based difficulties, drought, climatic-variability, poor physical health conditions, and past-injuries, the mental health of farmers get affected negativity. Studies on the farmers of both developing as well as developed countries have revealed of them to be experiencing worsened psychological health-disturbances that has pulled the policy-needs for considering how best to address different risk-factors incurring in the agri-sectors. Although the knowledge of these risk-factors that affect farmers' well-being is essential to reduce the work-related burdens on them, but implementation of technology-based solutions in this regard is very much essential for making the agriculture more profitable and sustainable.

References

Ainsworth, E. A., & Ort, D. R. (2010). How do we improve crop production in a warming world? *Plant Physiology, 154*, 526–530. http://www.ncbi.nlm.nih.gov

Altieri, M. A., & Nicholls, C. I. (2017). The adaptation and mitigation potential of traditional agriculture in a changing climate. *Climatic Change, 140*(1), 33–45.

Altieri, M. A., Nicholls, C. I., Henao, A., et al. (2015). Agroecology and the design of climate change-resilient farming systems. *Agronomy for Sustainable Development, 35*(3), 869–890.

Andrade, S. B., Anneberg, I., & Andrade, S. B. (2014). Farmers under pressure: Analysis of the social conditions of cases of animal neglect. *Journal of Agricultural and Environmental Ethics, 27*(1), 103–126.

Arizpe, N., Giampietro, M., & Ramos-Martin, J. (2011). Food security and fossil energy dependence: An international comparison of the use of fossil energy in agriculture (1991-2003). *Critical Reviews in Plant Sciences, 30*, 45–63.

Astrand, P. O., & Rodahl, K. (Eds.). (2003). *Textbook of work physiology: Physiological bases of exercise* (4th ed.). McGraw-Hill.

Backlund, P., Schimel, D., Janetos, A., et al. (2008). *Introduction. The effects of climate change on agriculture, land resources, water resources, and biodiversity in the United States* (p. 362). A Report by the U.S. Climate Change Science Program and the Subcommittee on Global Change Research.

Bates BC, Kundzewicz ZW, Wu S, et al. (2008) *Climate change and water*. Technical Paper of the Intergovernmental Panel on Climate Change. IPCC Secretariat, Geneva:210.

Bebber, D. P., Ramotowski, M. A., & Gurr, S. J. (2013). Crop pests and pathogens move polewards in a warming world. *Nature Climate Change, 3*, 985–988.

Bengtsson, J., Ahnström, J., & Weibull, A. C. (2005). The effects of organic agriculture on biodiversity and abundance: A meta-analysis. *Journal of Applied Ecology, 42*(2), 261–269.

Bhullar, N. K., & Gruissem, W. (2013). Nutritional enhancement of rice for human health: The contribution of biotechnology. *Biotechnology Advances, 31*(1), 50–57.

Bogue, R., & Bogue, R. (2016). Robots poised to revolutionise agriculture. *Industrial Robot, 43*(5), 450–456.

Cameron D, Osborne C, Horton P, et al. (2015) *A sustainable model for intensive agriculture*. Grantham Centre briefing note, The University of Sheffield 2015:4.

Carvajal, S. C., et al. (2014). Stress and sociocultural factors related to health status among US-Mexico border farmworkers. *Journal of Immigrant and Minority Health, 16*(6), 1176–1182.

Centre for Applied Psychological Research. (2012). *The Australian workplace barometer: Report on psychosocial safety climate and worker health in Australia*. University of South Australia.

Challinor, A. J., Watson, J., Lobell, D. B., et al. (2014). A meta-analysis of crop yield under climate change and adaptation. *Nature Climate Change, 4*, 287–291.

Chaney, B. H., & Torres, E. (2017). Covariates of identified stress and depression among seasonal farmworkers. *International Journal of Environmental Research and Public Health, 14*(7).

Cheng KH (2014) *Is vertical farming a more sustainable alternative to conventional farming in Hong Kong?* HKU Theses Online (HKUTO) 2014.

Dalin, C., Wada, Y., Kastner, T., et al. (2017). Groundwater depletion embedded in international food trade. *Nature, 543*(7647), 700–704.

De Castro, A. B., Krenz, J., & Neitzel, R. L. (2014). Assessing Hmong farmers' safety and health. *Workplace Health & Safety, 62*, 178–185.

Demos, K., Sazakli, E., Jelastopulu, E., Charokopos, N., Ellul, J., & Leotsinidis, M. (2013). Does farming have an effect on health status? A comparison study in West Greece. *International Journal of Environmental Research and Public Health, 10*(3), 776–792. https://doi.org/10.3390/ijerph10030776

Drinkwater, L. E., & Snapp, S. S. (2007). Nutrients in agroecosystems: Rethinking the management paradigme. *Advances in Agronomy, 92*. https://doi.org/10.1016/S0065-2113(04)92003-2

Dubowitz, T., Heron, M., Bird, C. E., et al. (2008). Neighborhood socioeconomic status and fruit and vegetable intake among whites, blacks, and Mexican Americans in the United States. *The American Journal of Clinical Nutrition, 87*(6), 1883–1891.

Erisman, J. W., Bleeker, A., Hensen, A., et al. (2008). Agricultural air quality in Europe and the future perspectives. *Atmospheric Environment, 42*(14), 3209–3217.

FAO (2010) *Climate smart agriculture: policies, practices and financing for food security, adaptation and mitigation.* Food and Agriculture Organization of the United Nations (FAO), Rome, Italy.

FAO. (2011a). *The state of the world's land and water resources for food and agriculture.* Managing systems at risk.

FAO (2011b) *The State of Food and Agriculture 2010–11: women in agriculture.* Closing the gender gap for development. Rome.

FAO (2012) *Developing a climate-smart agriculture strategy at the country-level: lessons from recent experience.* Background paper for the Second Global Conference on Agriculture, Food Security and Climate Change. Food and Agriculture Organization of the United Nations (FAO), Rome, Italy.

Fisher, J. R., Wynter, K. H., & Rowe, H. J. (2010). Innovative psycho-educational program to prevent common postpartum mental disorders in primiparous women: A before and after controlled study. *BMC Public Health, 2010*, 10. https://doi.org/10.1186/1471-2458-10-432

Flood, J. (2010). The importance of plant health to food security. *Food Security, 2*(3), 215–231.

Fraser, C. E., Smith, K. B., Judd, F., Humphreys, J. S., Fragar, L. J., & Henderson, A. (2005). Farming and mental health problems and mental illness. *The International Journal of Social Psychiatry, 51*(4), 340–349. https://doi.org/10.1177/0020764005060844

Gibbs, H. K., & Salmon, J. M. (2015). Mapping the world's degraded lands. *Applied Geography, 57*, 12–21.

Gomiero, T., Pimentel, D., & Paoletti, M. G. (2011). Is there a need for a more sustainable agriculture? *Critical Reviews in Plant Sciences, 30*, 6–23.

Gorgievski-Duijvesteijn, M., Giesen, C., & Bakker, A. (2000). Financial problems and health complaints among farm couples: Results of a 10-year follow-up study. *Journal of Occupational Health Psychology, 5*(3), 359–373. https://doi.org/10.1037//1076-8998.5.3.359

Goulson, D., Nicholls, E., Botías, C., et al. (2015). Bee declines driven by combined stress from parasites, pesticides, and lack of flowers. *Science, 347*(6229), 1255957.

Green, R. E., Cornell, S. J., Scharlemann, J. P. W., & Balmford, A. (2005). Farming and the fate of wild nature. *Science, 307*, 550–555. https://doi.org/10.1126/science.1106049

Hagen, B. N. M., Albright, A., Sargeant, J., Winder, C. B., Harper, S. L., O'Sullivan, T. L., & Jones-Bitton, A. (2019). Research trends in farmers' mental health: A scoping review of mental health outcomes and interventions among farming populations worldwide. *PLoS One, 14*(12), e0225661. https://doi.org/10.1371/journal.pone.0225661

Hanlon, C., Whitley, R., Wondimagegn, D., Alem, A., & Prince, M. (2010). Between life and death: Exploring the sociocultural context of antenatal mental distress in rural Ethiopia. *Archives of Women's Mental Health, 13*, 385–393.

Hu, H., & Xiong, L. (2014). Genetic engineering and breeding of drought-resistant crops. *Annual Review of Plant Biology, 65*, 715–741.

IPCC: Climate Change. (2014). In C. B. Field, V. R. Barros, D. J. Dokken, et al. (Eds.), *Impacts, adaptation, and vulnerability. Part A: Global and sectoral aspects. Contribution of working group II to the fifth assessment report of the intergovernmental panel on climate change* (p. 2014).

Jenkins, R., Njenga, F., Okonji, M., Kigamwa, P., Baraza, M., Ayuyo, J., & Kiima, D. (2012). Prevalence of common mental disorders in a rural district of Kenya, and socio-demographic risk factors. *International Journal of Environmental Research and Public Health, 9*, 1810–1819.

Kilkkinen, A., Kao-Philpot, A., O'Neil, A., Philpot, B., Reddy, P., Bunker, S., & Dunbar, J. (2007). Prevalence of psychological distress, anxiety and depression in rural communities in Australia. *The Australian Journal of Rural Health, 15*, 114–119.

Klimont, Z., Smith, S. J., & Cofala, J. (2013). The last decade of global anthropogenic sulfur dioxide: 2000–2011 emissions. *Environmental Research Letters, 8*(1), 014003.

Koohafkan, P., Altieri, M. A., & Gimenez, E. H. (2012). Green agriculture: Foundations for biodiverse, resilient and productive agricultural systems. *International Journal of Agricultural Sustainability, 10*(1), 61–75.

Krantz, G., Forsman, M., & Lundberg, U. (2004). Consistency in physiological stress responses and electromyography activity during induced stress exposure in women and men. *Integrative Physiological & Behavioral Science, 39*(2), 105–118.

Laurance, W. F., Sayer, J., & Cassman, K. G. (2014). Agricultural expansion and its impacts on tropical nature. *Trends in Ecology & Evolution, 29*(2), 107–116.

Lin, B. B. (2011). Resilience in agriculture through crop diversification: Adaptive management for environmental change. *Bioscience, 61*(3), 183–193.

Linn, J. G., & Husaini, B. A. (1987). Determinants of psychological depression and coping behaviors of Tennessee farm residents. *Journal of Community Psychology, 15*(4), 503–512.

Lipscomb, H. J., Dement, J. M., Epling, C. A., Gaynes, B. N., McDonald, M. A., & Schoenfisch, A. L. (2007). Depressive symptoms among working women in rural North Carolina: A comparison of women in poultry processing and other low-wage jobs. *International Journal of Law and Psychiatry, 30*, 284–298.

Lobell, D. B., & Gourdji, S. M. (2012). The influence of climate change on global crop productivity. *Plant Physiology, 160*(4), 1686–1697.

Ma Elena, C.-J. (2009). Health consequences of rural Women's productive role in agriculture in the Philippines and other developing countries. *Journal of International Women's Studies, 10*(4), 95–110. Available at: http://vc.bridgew.edu/jiws/vol10/iss4/7

McIntyre B, Herren H, Wakhungu J, Watson R (2009) *Agriculture at a crossroads: International assessment of agricultural knowledge*.

Meldrum, G., Sthapit, S., Rojas, W., et al. (2013). Agricultural biodiversity enhances capacity to adapt to climate change. *New Agriculturist, 2013*, 13.

Miller, V., Yusuf, S., Chow, C. K., et al. (2016). Availability, affordability, and consumption of fruits and vegetables in 18 countries across income levels: Findings from the prospective urban rural epidemiology (PURE) study. *The Lancet Global Health, 4*(10), e695–e703.

Olson, K. R., & Schellenberg, R. P. (1986). Farm stressors. *American Journal of Community Psychology, 14*(5), 555–569. https://doi.org/10.1007/bf00935358

Pessoa, M. C., Mendes, L. L., Gomes, C. S., et al. (2015). Food environment and fruit and vegetable intake in a urban population: A multilevel analysis. *BMC Public Health, 15*, 1012.

Pichorner, A., Schaffernicht, S., Schick, M., Moder, K., & Quendler, E. (2017). Working time requirements, workload and physical strain of daily stable work on small farms specialized in suckler cows in upper Carinthia. *Chemical Engineering Transactions, 58*, 43–48. https://doi.org/10.3303/CET1758008

Porter, J. R., Xie, L., Challinor, A. J., et al. (2014). Food security and food production systems. In C. B. Field, V. R. Barros, D. J. Dokken, et al. (Eds.), *Climate change 2014: Impacts, adaptation, and vulnerability. Part a: Global and sectoral aspects. Contribution of working group II to the fifth assessment report of the intergovernmental panel on climate change* (pp. 485–533). Cambridge University Press.

Pretty, J. (2008). Agricultural sustainability: Concepts, principles and evidence. *Philosophical Transactions of the Royal Society B, 363*, 447–465. https://doi.org/10.1098/rstb.2007.2163

Randall B, Jeter G (2003) *A guide to manual materials handling and back safety*. N.C. Department of Labor, Division of Occupational Safety and Health. Retrieved from http://www.nclabor.com/osha/etta/indguide/ig26.pdf

Savary, S. (2014). The roots of crop health: Cropping practices and disease management. *Food Security, 6*, 819–831. https://doi.org/10.1007/s12571-014-0399-4

Shidhaye, R., & Patel, V. (2010). Association of socio-economic, gender and health factors with common mental disorders in women: A population-based study of 5703 married rural women in India. *International Journal of Epidemiology, 39*, 1510–1521.

Smith, P., Bustamante, M., Ahammad, H., et al. (2014). Agriculture, forestry and other land use (AFOLU). In O. Edenhofer, R. Pichs-Madruga, & Y. Sokona (Eds.), *Climate change 2014: Mitigation of climate change. Contribution of working group III to the fifth assessment report of the intergovernmental panel on climate change*. Cambridge University Press.

Smith, M. R., Singh, G. M., Mozaffarian, D., et al. (2015). Effects of decreases of animal pollinators on human nutrition and global health: A modelling analysis. *Lancet, 386*(10007), 1964–1972.

Tilman, D., Fargione, J., Wolff, B., D'Antonio, C., Dobson, A., Howarth, R., Schindler, D., Schlesinger, W. H., Simberloff, D., & Swackhamer, D. (2001). Forecasting agriculturally driven global environmental change. *Science, 292*, 281–284.

Tsao, Y., Creedy, D. K., & Gamble, J. (2015). Prevalence and psychological correlates of postnatal depression in rural Taiwanese women. *Health Care for Women International, 36*, 457–474.

Tuomisto, H. L., Hodge, I. D., Riordan, P., et al. (2012). Does organic farming reduce environmental impacts? –a meta-analysis of European research. *Journal of Environmental Management, 112*, 309–320.

Tuomisto, H. L., Scheelbeek, P. F. D., Chalabi, Z., Green, R., Smith, R. D., Haines, A., & Dangour, A. D. (2017). Effects of environmental change on agriculture, nutrition and health: A framework with a focus on fruits and vegetables. *Wellcome Open Research, 2*, 21. https://doi.org/10.12688/wellcomeopenres.11190.2

Vink, P., & Hallbeck, S. (2012). Editorial: Comfort and discomfort studies demonstrate the need for a new model. *Applied Ergonomics, 43*, 271–276.

Vyas, R. (2012). Mitigation of musculoskeletal problems and body discomfort of agricultural workers through educational intervention. *Work, 41*, 2398–2404.

WHO. (1992). *The ICD-10 classification of mental and Behavioural disorders*. Clinical Descriptions and Diagnostic Guidelines; World Health Organization.

WHO. (2001). *Mental health: New understanding, new hope*. World Health Organization.

WHO. (2013a). *Mental health action plan 2013–2020*. World Health Organization.

WHO. (2013b). *WHO global plan of action on workers' health (2008–2017): Baseline for implementation. Global country survey 2008/2009*. Executive Summary and Survey Findings; World Health Organization.

Wilhelmsson, S., Andersson, M., Arvidsson, I., Dahlqvist, C., Hemsworth, P. H., Yngvesson, J., & Hultgren, J. (2021). Physical workload and psychosocial working conditions in Swedish pig transport drivers. *International Journal of Industrial Ergonomics, 83*, 103124.

Yazd, S. D., Wheeler, S. A., & Zuo, A. (2019). Key risk factors affecting farmers' mental health: A systematic review. *International Journal of Environmental Research and Public Health, 16*, 4849. https://doi.org/10.3390/ijerph16234849

Zhang, C., & Kovacs, J. M. (2012). The application of small unmanned aerial systems for precision agriculture: A review. *Precision Agriculture, 13*(6), 693–712.

Chapter 2
Literature on the Global Agri-Sectors: An Overview

2.1 Introduction

Both from urban and rural areas, still the labor is attracted towards agriculture which has been reported in providing employments for "10 million-people in 2005" and "15 million-people in 2010" (Herderschee et al., 2012). With the aim of obtaining necessary food for the subsistence, the labor and small-scale farmers get attracted to agriculture despite of lower productivity in most cases. The worker's workload results through the interaction arising from inherent-demands to environment-conditions and bio-psychological capacity of human-being, which can affect continuum of human health-related diseases leading to disorder occurring psychologically and physically (SECCO et al., 2011). Further, the workloads can result in stresses, fatigue, occupational-accidents and diseases (Hart, 2006). Thus, it becomes essential in understanding of the agricultural-work and its influences on workers' health in order to the existing linkage among work, health, and associated diseases in this context. With regard to the mental-stress, a higher prevalence of mental-disorders occurs due to stress, including anxiety and depressions (McShane et al., 2016), which arises owing to higher work-demands that the workers become incapable to cope, manage and control (Lunner Kolstrup et al., 2013).

The farmers' roles become more stressful with the advent of future climatic changes (Kearney et al., 2014). In a study, Williams (2001) has reported of chronic-stress leading to physical-problems and mental-problems among farm workers that may be further augmented with cognitive-issues like memory-losses and inability in making decisions. Thus, addressing the existing farming practices throughout the world that have been posing the global farming communities to work as well as mental related stresses become a serious concern in the recent days.

S. Satapathy et al., *Innovation in Agriculture with IoT and AI*,
Innovation, Technology, and Knowledge Management,
https://doi.org/10.1007/978-3-030-88828-2_2

2.2 Literature Review on Agri-Farming

There has been considerable development in the agri-farming sectors with more utilization of agricultural lands in the recent years for making "export-oriented and large-scale commercial agriculture", where the agriculture-based investments are made by foreign-investors for the purpose of securing their food-needs. This, in-turn, constrains the small-scale farmers to access the lands for use. The enhancements in agricultural productivity have been a fundamental-policy for the initiation of transformation in agriculture with significant increase in poorer income-levels (Ngai & Pissarides, 2007). Studies have revealed of stronger as well as robust effects on poverty-levels by agriculture productivity (de Janvry & Sadoulet, 2010; Irz et al., 2001; Thirtle et al., 2003). However, depending upon the country and the development as well as use of newer technologies, the magnitude of poverty-reduction varies with respect to the growth in agricultural productivity (de Janvry & Sadoulet, 2010).

The technology-based changes have been regarded to be influencing the growth in productivity (DFID, 2006; Morris et al., 2007; OECD, 2012). However, it is worth mentioning that the innovation, selection, and adoption of new technologies depend on the agriculture frontier, factor endowment, and market imperfections. Hayami and Ruttan (1970) have analyzed the technology-based changes in Japan and the US during '1880 to 1960' and it was found that in the US, the abundance of land favored 'labor-saving' technology-based changes. On the other hand, the scarcity of land in Japan resulted in developing and adopting 'land-saving' technologies. The "Green Revolution (GR)" in Asia was an augmentative input-oriented production using new varieties of seeds such as wheat and rice that can provide high-yield and were fertilizer-efficient (Eicher, 1995; Hazell, 2009; Smale, 1995). This was further initiated by the policy makers for significant transformations in agriculture for increasing food-production and for reducing malnutrition as well as hunger in the 1960s. However, different arguments state that the GR has lessened income-distribution by missing of the small-scale farmers and poorer subsistence, while favoring to the larger farmers (Das, 1998; Freebairn, 1995; Goldman & Smith, 1995; Griffifin, 1979). Further, it also increased the numbers of landless-farmers in addition to unskilled labor's demand by lowering wage-laborers (Cleaver Jr, 1972; Glaeser, 1987; Hazell & Ramasamy, 1991). The agriculture-based mechanization has been considered in reducing the costs of production and improving the productivity in addition to taking of quality as well as timely agricultural-operations. However, owing to a number of significant constraints, the non-use of modern farm-technology occur in agriculture (Pattnaik & Mohapatra, 2019).

2.3 Literature Review on Physical Workload

Significant losses of working-hours occur due to serious injuries or illness, and this substantially causes health-care costs. Approximately 370-million US Dollars have been received in 2014 as disability pension in Switzerland, where primary reasons of disability were revealed as illness followed by musculoskeletal-disorders (Federal Office of Social Insurances, 2015). The physical workload is usually associated with different aspects, including heavy-lifting tasks, awkward-postures, repetitive-work, and lack of recuperation (Arvidsson et al., 2012; da Costa & Vieira, 2010). With long-exposures to higher physical-load, the occurrence of muscular-disorders become predominant, resulting in back as well as shoulder pains (Dalbøge et al., 2014), which requires proper technical-recordings for the associated ergonomic-risks' evaluation (Balogh et al., 2019; Hansson et al., 2001; Hansson et al., 2010). The cause of musculoskeletal-disorders is mainly due to the risk factor of higher physical workload (Karpansalo et al., 2002). Usually, the work-capacities of individuals are checked for their job-profile adjustments (Soer et al., 2009). Thus, the knowledge of workload as well as work-capacity can help in facilitating successful work-resumption (Soer et al., 2009). The use of "functional-capacity evaluations" has been recommended to assess the work-capacity (Soer et al., 2008). Soer et al. (2009) assessed the work-capacities of employees by using 12 work-related tests. However, many studies have measured the physical-activity in employees using accelerometer or pedometer in addition to questionnaires of self-reported (Miller & Brown, 2004; Steele & Mummery, 2003; Thorp et al., 2012; Tudor-Locke et al., 2013). Some researchers have used armband for the detection of energy-expenditure in physical-activities (Brighenti-Zogg et al., 2016; Johannsen et al., 2010). The individual's states of stresses comprise of two components, such as the stressors (demand-source) and the stress-responses (e.g., physiological, affective, and cognitive elements). In some cases, the stress-responses may involve worry-levels compromising the person's ability to get panic as well as tension, lacking concentration, feeling of apprehensions, and physiological-reactions, including shallow-breathing, augmented heart-rate, perspiration and tense-muscles (Lazarus, 1999). However, the stressors may remain the same for two individuals, but they may experience with different stress-levels.

Rocha et al. (2016) have studied a total of 259 farm workers from two rural-environments, and found the male as well as female farm-workers of one-environment reporting of the total-effort's demand contributed in workloads, while the female farm-workers of the other-environment reported of physical-demand to be contributing in workloads. There has been insufficient safety-management with regard to pig's transport from farmers (Valluru et al., 2017), which violates the best work-practices. In the pig-production related farming, the animal's handling has been a major cause for the occurrences of handlers' physical injuries (Langley & Morrow, 2010), with frequent utilization of handling-tools during pig's unloading activities (Bornhede, 2014). The employees engaged in transportation activities are frequently reported of work-related injuries and accidents (Chandler et al., 2017;

Wiatrowski & Janocha, 2014), with majority of accidents/injuries based cases to be occurring while goods' handling outside the truck (Chandler et al., 2017) and also because of the associated psycho-social risk-factors (European Agency for Safety and Health at Work, 2011).

2.4 Literature Review on Mental Workload

A number of occupational health-risks have been reported by different researchers with some studies specifying the farming to be more stressful-occupation than others (Fraser et al., 2005; Gregoire, 2002; Price & Evans, 2009; Roy et al., 2013). Moreover, farming is related to different physical as well as mental-health risks owing to its harder tasks under different challenging-conditions (Fraser et al., 2005). The mental-stress or mental ill-health has been defined to be combining some sort of emotions, abnormal-thoughts, behaviour, and associations with others (World Health Organization, 2007a), which includes depressions, anxieties, stresses, bipolar-disorder, schizophrenia, and psychological or emotional distresses (World Health Organization, 2007b). Moreover, it continues to grow with substantive health related impacts with major and globally "human-rights' related, social-related, and economic-related" consequences (World Health Organization, 2007b). Past studies have identified associated common risk-factors influencing farmers' mental-health around the world and having a major influence on their well-being as well as health, which included commodity-prices, debt, climate-changes, drought, over-work, government-regulations, working isolated, role-conflicts, time-related work-pressures, and poor-housing conditions (Bin et al., 2008; Carvajal et al., 2014; Deary et al., 1997; Firth et al., 2007; Grzywacz et al., 2010; Hanklang et al., 2016; Hedlund & Berkowitz, 1979; Hiott et al., 2008; Hossain et al., 2008; Hovey & Magaña, 2002a, 2002b; Logstein, 2016; McShane et al., 2016; Mora et al., 2016; Raine, 1999; Ramos et al., 2015; Simkin et al., 1998; Staniford et al., 2009; Vega et al., 1985; Walker et al., 1986; Wheeler et al., 2018; Williams, 1996).

The impact of mental ill-health among farm workers can affect their normal-lives in varying ways including: less-interest in pleasure, lesser concentrations, loss of appetite, weight changes, irritability, tiredness, sleeping-related problems, anxieties, fatigue, loss of self-control, self-esteem losses, retraction from social activity, relationship's break-down, memory-losses, lack of tempers, easiness related problems, blue-feeling, and substance abuse, respectively (Glasscock et al., 2006; Liu et al., 2018; O'Connor et al., 2015; Olowogbon et al., 2019; Simsek et al., 2015; Terrazas & McCormick, 2018; Walker & Walker, 1987). Also in several studies, higher suicide-rates among the farm-based individuals owing to their mental ill-health conditions have been reported (Das, 2011; Kunde et al., 2017; Liu et al., 2018; Page & Fragar, 2002; Perceval et al., 2017), which has been a serious-concern affecting the farming-communities throughout the world. Considerable association between the mental health-illness among the farmers and their exposures to pesticides have been reported (Aiwerasia et al., 2001; Corral et al., 2017; Jamal et al.,

2002; Malekirad et al., 2013; Muñoz-Quezada et al., 2017; Steenland et al., 1994) and also, depression-rates to be higher with pesticide-exposures (Atreya et al., 2012; Bayrami et al., 2012; Beseler, 2005; Conti et al., 2018; Da Silva et al., 2016; Fariba et al., 2016; Harrison & Mackenzie Ross, 2016; Hong et al., 2008; Hong et al., 2009; Kim et al., 2013; Povey et al., 2014; Quandt et al., 2010; Siegel et al., 2017; Suten Geofrey et al., 2018; Weisskopf et al., 2013; Wesseling et al., 2010; Zhang et al., 2016). Some studies have revealed of the finance-based challenges impacting the mental health-conditions of farmers (Bryant & Garnham, 2014; Eberhardt & Pooyan, 1990; Feng et al., 2015; Hovey & Magaña, 2002a, 2002b; Kallioniemi et al., 2016; Kureshi & Somsundaram, 2018; Pulgar et al., 2016; Ramos et al., 2015; Simsek et al., 2015; Truchot & Andela, 2018; Vega et al., 1985; Williams, 1996). Bultena et al. (1986) have reported of farmers with psychological-distress, depressions, reduced life-satisfaction, addiction to alcohols, and even attempting to suicides. Earlier studies have found of positive relationship between higher farm-profits, greater well-beings, and lesser distress among the farm-related workers (Peel et al., 2015). Based on the reporting of Walker et al. (1986), the unfavorable climatic-conditions and the weather's unpredictability were the major key-stresses in "North America". Further, Kearney et al. (2014) have found of a total 60% farmers in "Eastern North Carolina" reporting of more stressful situation by working for more than 40 hours-per-week in bad weather-conditions. In a study, Edwards et al. (2015) have found of more adverse impacts on the mental conditions of farmers owing to the severity of drought. In addition, the mental conditions of farmers can be greatly affected with the summer heat waves as the farm workers often have no choices left other than but working in extreme hot weather conditions (Singh et al., 2013). Moreover, in different studies owing to poor physical health, and past-injuries or work-disabilities, a higher rate of mental health-illness among the farm workers have been reported (Çakmur, 2014; Kallioniemi et al., 2009; Linn & Husaini, 1987; Onwuameze et al., 2013).

2.5 Literature Review on Agri-Problems and Models

The world has been occupied by more than 7-billion people with an expectation of further growth to 9-billion by the year '2050' (Simmons, 2011), which implied for the requirement of more agriculture-based advancements in order to feed many people with safe and nutritional food. Over 200 years-ago in the United States, 90% of the population were living on farms by producing their consumable food. Based on the statement by Prax (2010), only a two percent mass of now-a-days are producing food including vegetables, fruits, diary and meat. The adoption of technology by the farmers is to make advances to produce more foods. At the present scenario, a farmer is able to feed "155 individuals" owing to technological adoptions, but in the past, a farmer could be able to feed only "19 individuals" during 1940s (Prax, 2010). Sahu and Mishra (2014), in their analysis of various socioeconomic factors that influenced farmer's technology-adaptability, found the household-size to be influenced with the adapting-decision. However, Tizale (2007) has stated that, in order to help in the consumption-pattern of larger families, a part of family-members get involved in the off-farm sectors. In a study in "South Western Nigeria", Saka et al. (2005) have found significant-adoption of improved rice-varieties among the small-holder farmers. Mendis and Udomsade (2005) have identified the factors relating to farmers' crop-management adoption-practices as their education-level, land-size, incomes, credits, information-sources, land-tenure, engagement in extension-activities, visits to extension-offices, and membership in the organizations for farmers.

In view of the associated problems in farming, it has been an important concern to improve the mental-health as well as well-being of the whole farming-communities by enhancing the required resilience for managing and overcoming of uncertainties (Davies et al., 2019). The possible agricultural uncertainties and challenges that have adverse impacts on mental-health and well-being of farmers, and suitable preventive as well as promotional activities to support resilience in this aspect are shown in Fig. 2.1.

Tuomisto et al. (2017) have presented an overall-framework in context to the farmers' socio-economic and societal factors including "culture, religion, wealth-distribution, and population-structure" for the agricultural effects of environmental changes (Fig. 2.2). The other five factors that were considered in this framework included "environmental changes: stressors affecting directly the food-systems", "interventions and policies: research and innovation; technological development; and government policies", "food-system activities: production of inputs; infrastructure; agricultural processes; food processing, trade and consumption; and waste

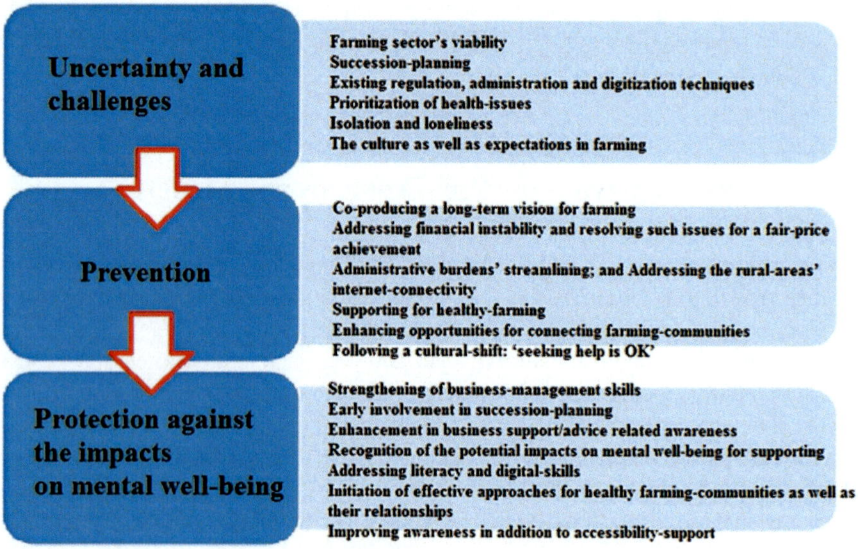

Fig. 2.1 Prevention of uncertainties and challenges in agriculture, and protection against the impacts on mental well-being of farmers

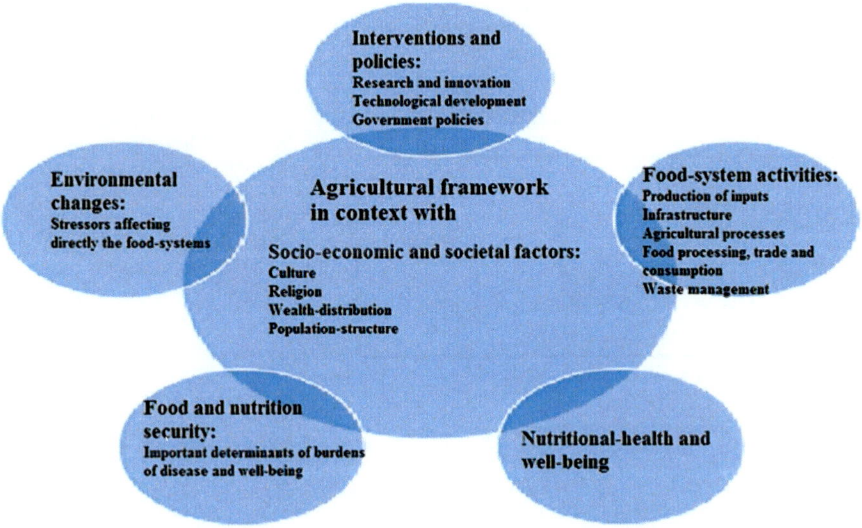

Fig. 2.2 An overall framework for the agricultural effects of environmental changes

management", "food and nutrition security: important determinants of burdens of disease and well-being", and nutritional-health and well-being, respectively.

The technology-adoption concept in agricultural research is useful in designing and developing interventions, framing of evaluations' impact and in informing decision-making regarding newer investments in development-based works. Glover et al. (2019) have characterized the technology-change programme into four-aspects, such as "propositions; encounters; dispositions; and responses"; respectively. In the global South, the development and spreading of new-technology has been considered as a major concern to enhance the resilience, productivity and sustainability of small-scale farming-systems (InterAcademy Council, 2004; McIntyre et al., 2009). There are some studies that focused in the theory of technological adoption as well as diffusion, adoption-limitations, and their measuring ways (Doss, 2006; Rogers, 2003; Temple et al., 2016).

For any development-based agricultural research, the processes involved in technological changes and their outcomes assessments are prevailing challenges to the involved people as well as organizations. The technological "generation, testing, and dissemination" are of prime importance for any agricultural development (Thornton et al., 2017). The present day technological advancements in agriculture with the implementation of "Internet of Things (IoT)" in farm's record keeping as well as monitoring in order to accomplish the objectives in agriculture through wireless network is illustrated in Fig. 2.3, and it has also resulted in higher agricultural performances and outcomes in comparison to the conventional techniques. The IoT-based network has been more beneficial and useful for the agricultural sectors (Farooq et al., 2019), which helps in monitoring of farm related data and help in facilitating the transmission as well as reception of various useful data and

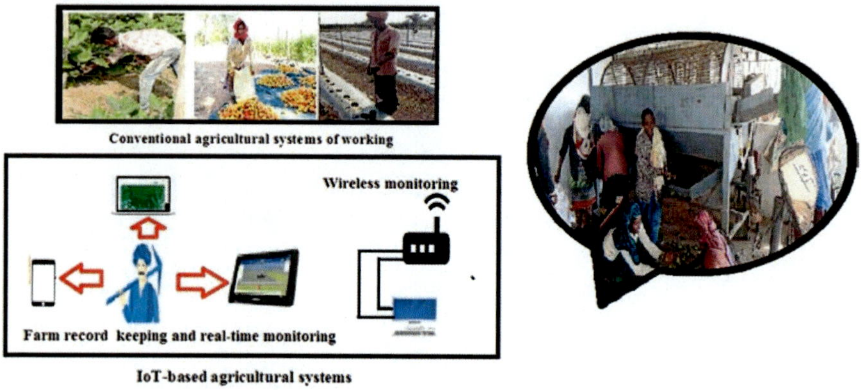

Fig. 2.3 Technological advancement in agriculture

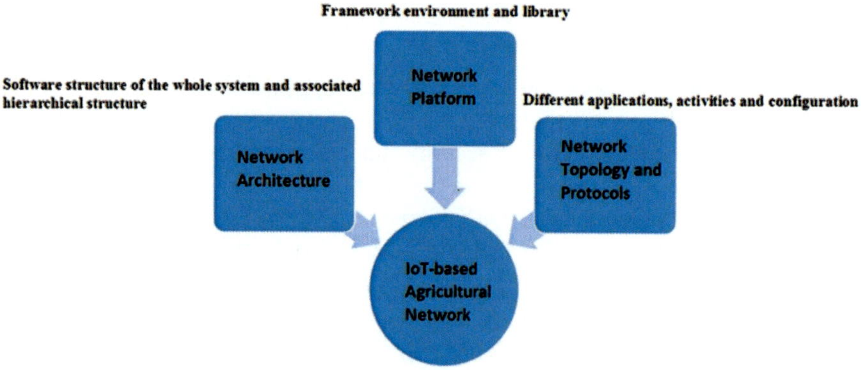

Fig. 2.4 IoT-based network for agricultural sectors

information. As shown in Fig. 2.4, the IoT-based network's framework comprises of network architecture, network platform, and network topology as well as protocols.

For getting information on agriculture and to collect different agricultural data that has become a complicated task in conventional farming. Then, the use of different sensors as well as monitoring devices for sensing and monitoring can be more viable. The use of sensors can also be useful in collecting data with subsequent storage of such data on crop-diseases. Moreover, the farmers can be able to obtain necessary information regarding "temperature, PH value, soil-moisture and humidity" through message as well as web services. The communication-protocols have been considered to be important elements in IoT-based agriculture systems, which include the internet-related technologies like "WIFI, ZigBee, LoraWan, and Code-Division Multiple-Access (CDMA)".

Further, the use of micro-controller to perform the operations related to networking in agricultural applications have been increasing these days. Based on the functional-framework suggested by Farooq et al. (2019), Fig. 2.5 illustrates an

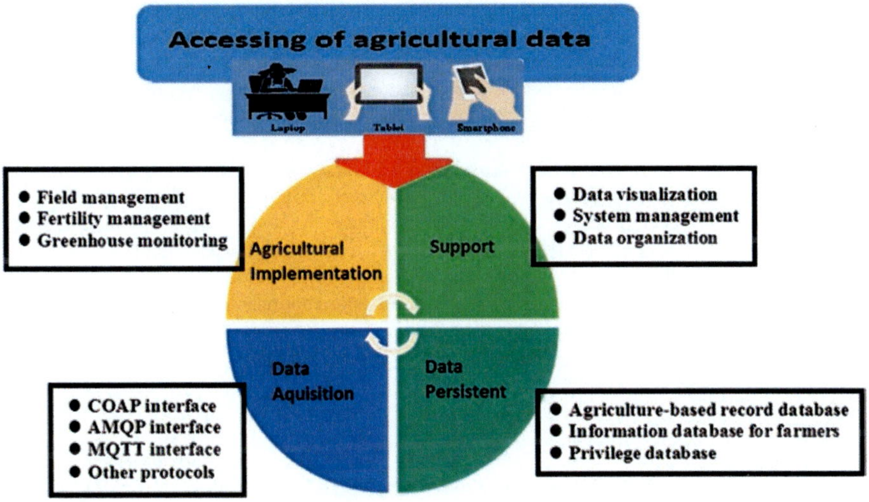

Fig. 2.5 IoT-based functional-framework for information-service model in agriculture

IoT-based information-service model in agriculture. The functional framework shows how agriculturists and farmers can access multiple databases with the assistance of support layer from application layer. Business layer contains all necessary operation which are important for any IoT farm. Data acquisition layer provides connection with session layer through IoT protocols like MQTT, AMQP and COAP etc.

References

Aiwerasia, V. F. N., David, N. M., Timo, J. P., Michael, P. S., & Godson, M. (2001). Acute health effects of organophosphorus pesticides on Tanzanian small-scale coffee growers. *Journal of Exposure Analysis and Environmental Epidemiology, 11*, 335.

Arvidsson, I., Balogh, I., Hansson, G. A., Ohlsson, K., Akesson, I., & Nordander, C. (2012). Rationalization in meat cutting - consequences on physical workload. *Applied Ergonomics, 43*(6), 1026–1032. https://doi.org/10.1016/j.apergo.2012.03.001

Atreya, K., Kumar Sitaula, B., Overgaard, H., Man Bajracharya, R., & Sharma, S. (2012). Knowledge, attitude and practices of pesticide use and acetylcholinesterase depression among farm workers in Nepal. *International Journal of Environmental Health Research, 22*, 401–415.

Balogh, I., Arvidsson, I., Bjork, J., Hansson, G. A., Ohlsson, K., Skerfving, S., & Nordander, C. (2019). Work-related neck and upper limb disorders - quantitative exposure-response relationships adjusted for personal characteristics and psychosocial conditions. *BMC Musculoskeletal Disorders, 20*(1), 139. https://doi.org/10.1186/s12891-019-2491-6

Bayrami, M., Hashemi, T., Malekirad, A. A., Ashayeri, H., Faraji, F., & Abdollahi, M. (2012). Electroencephalogram, cognitive state, psychological disorders, clinical symptom, and oxidative stress in horticulture farmers exposed to organophosphate pesticides. *Toxicology and Industrial Health, 28*, 90–96.

Beseler C (2005) *Diagnosed depression and low, intermediate, and high pesticide exposures in Iowa and North Carolina farm applicators and their spouses enrolled in the agricultural health study*. Ph.D. Thesis, Colorado State University, Fort Collins, CO, USA.

Bin, H., Lamm, F., & Tipples, R. (2008). The impact of stressors on the psychological wellbeing of New Zealand farmers and the development of an explanatory conceptual model. *Policy and Practice in Health and Safety, 6*, 79–96.

Bornhede, M. (2014). *A comparison of transporters' paddle use when unloading pigs at slaughter*. Master Thesis. Swedish University of Agricultural Sciences, Uppsala.

Brighenti-Zogg, S., Mundwiler, J., Schüpbach, U., Dieterle, T., Wolfer, D. P., Leuppi, J. D., & Miedinger, D. (2016). Physical workload and work capacity across occupational groups. *PLoS One, 11*(5), e0154073. https://doi.org/10.1371/journal.pone.0154073

Bryant, L., & Garnham, B. (2014). Economies, ethics and emotions: Farmer distress within the moral economy of agribusiness. *Journal of Rural Studies, 34*, 304–312.

Bultena, G., Lasley, P., & Geller, J. (1986). The farm crisis: Patterns and impacts of financial distress among Iowa farm families. *Rural Sociology, 51*, 436.

Çakmur, H. (2014). Health risks faced by Turkish agricultural workers. *Scientific World Journal, 2014*, 1–8.

Carvajal, S., Kibor, C., McClelland, D., Ingram, M., Zapien, J., Torres, E., Redondo, F., Rodriguez, K., Rubio-Goldsmith, R., Meister, J., et al. (2014). Stress and sociocultural factors related to health status among US–Mexico border farmworkers. *Journal of Immigrant and Minority Health, 16*, 1176–1182.

Chandler, M. D., Bunn, T. L., & Slavova, S. (2017). Narrative and quantitative analyses of workers' compensation-covered injuries in short-haul vs. long-haul trucking. *International Journal of Injury Control and Safety Promotion, 24*(1), 120–130. https://doi.org/10.1080/17457300.2016.1170041

Cleaver, H. M., Jr. (1972). The contradictions of the green revolution. *The American Economic Review, 62*(1/2), 177–186.

Conti, C. L., Barbosa, W. M., Simão, J. B. P., & Álvares-da-Silva, A. M. (2018). Pesticide exposure, tobacco use, poor self-perceived health and presence of chronic disease are determinants of depressive symptoms among coffee growers from Southeast Brazil. *Psychiatry Research, 260*, 187–192.

Corral, S. A., de Angel, V., Salas, N., Zúñiga-Venegas, L., Gaspar, P. A., & Pancetti, F. (2017). Cognitive impairment in agricultural workers and nearby residents exposed to pesticides in the Coquimbo Region of Chile. *Neurotoxicology and Teratology, 62*, 13–19.

Council, I. A. (2004). *Realizing the promise and potential of African agriculture*. InterAcademy Council.

da Costa, B. R., & Vieira, E. R. (2010). Risk factors for work-related musculoskeletal disorders: A systematic review of recent longitudinal studies. *American Journal of Industrial Medicine, 53*(3), 285–323. https://doi.org/10.1002/ajim.20750

Da Silva, V. S. P., de Mello, M. S. C., & Otero, U. B. (2016). Exposure to pesticides and mental disorders in a rural population of southern Brazil. *Neurotoxicology, 56*, 7–16.

Dalbøge, A., Frost, P., Andersen, J. H., & Svendsen, S. W. (2014). Cumulative occupational shoulder exposures and surgery for subacromial impingement syndrome: A nationwide Danish cohort study. *Occupational and Environmental Medicine, 71*(11), 750–756. https://doi.org/10.1136/oemed-2014-102161

Das, R. J. (1998). The green revolution, agrarian productivity and labor. *International Journal of Urban and Regional Research, 22*(1), 122–135. https://doi.org/10.1111/1468-2427.00127

Das, A. (2011). Farmers' suicide in India: Implications for public mental health. *The International Journal of Social Psychiatry, 57*, 21–29.

Davies, A. R., Homolova, L., Grey, C. N. B., Fisher, J., Burchett, N., & Kousoulis, A. (2019). *Supporting farming communities at times of uncertainty: An action framework to support the mental health and well-being of farmers and their families*. Public Health Wales NHS Trust & Mental Health Foundation.

de Janvry, A., & Sadoulet, E. (2010). Agricultural growth and poverty reduction: Additional evidence. *World Bank Research Observer, 25*(1), 1–20. https://doi.org/10.1093/wbro/lkp015

Deary, I. J., Willock, J., & McGregor, M. (1997). Stress in farming. *Stress Medicine, 13*, 131–136.

DFID. (2006). *Technology and its contribution to pro-poor agricultural development*. DFID working paper 4, University of Greenwich.

Doss, C. R. (2006). Analyzing technology adoption using micro studies: Limitations, challenges, and opportunities for improvement. *Agricultural Economics, 34*, 207–219.

Eberhardt, B. J., & Pooyan, A. (1990). Development of the farm stress survey: Factorial structure, reliability, and validity. *Educational and Psychological Measurement, 50*, 393–402.

Edwards, B., Gray, M., & Hunter, B. (2015). The impact of drought on mental health in rural and regional Australia. *Social Indicators Research, 121*, 177–194.

Eicher, C. K. (1995). Zimbabwe's maize-based Green revolution: Preconditions for replication. *World Development, 23*(5), 805–818. https://doi.org/10.1016/0305-750X(95)93983-R

European Agency for Safety and Health at Work. (2011). OSH in figures: Occupational safety and health in the transport sector – An overview. *Luxembourg*. https://doi.org/10.2802/2218. ISBN 978-92-9191-303-9.

Fariba, T., Gholamhassan, V., Mohammad, A., & Ali Akbar, M. (2016). A comparative study of the quality of life, depression, anxiety and stress in farmers exposed to organophosphate pesticides with those in a control group. *Journal of Chemical Health Risks, 6*, 143–151.

Farooq, M. S., Riaz, S., Abid, A., Abid, K., & Naeem, M. A. (2019). A survey on the role of IoT in agriculture for the implementation of SMART farming. Special section on new technologies for smart farming 4.0: Research challenges and opportunities. *IEEE, 7*, 156237–156271.

Federal Office of Social Insurances (2015) *Statistics of social security: IV-Statistics 2014*. Available: http://www.bfs.admin.ch/bfs/portal/de/index/news/publikationen.html?publicationID=6588. Accessed 29 July 2020.

Feng, D., Ji, L., & Xu, L. (2015). Effffect of subjective economic status on psychological distress among farmers and non-farmers of rural China. *The Australian Journal of Rural Health, 23*, 215–220.

Firth, H. M., Williams, S. M., Herbison, G. P., & McGee, R. O. (2007). Stress in New Zealand farmers. *Stress Health Journal of the International Society for the Investigation of Stress, 23*, 51–58.

Fraser, C. E., Smith, K. B., Judd, F., Humphreys, J. S., Fragar, L. J., & Henderson, A. (2005). Farming and mental health problems and mental illness. *The International Journal of Social Psychiatry, 51*, 340–349.

Freebairn, D. (1995). Did the green revolution concentrate incomes? A quantitative study of research reports. *World Development, 23*, 265–279. https://doi.org/10.1016/0305-750X(94)00116-G

Glaeser, B. (1987). *The Green revolution revisited*. Allen & Unwin.

Glasscock, D. J., Rasmussen, K., Carstensen, O., & Hansen, O. N. (2006). Psychosocial factors and safety behaviour as predictors of accidental work injuries in farming. *Work and Stress, 20*, 173–189.

Glover, D., Sumberg, J., Ton, G., Andersson, J., & Badstue, L. (2019). Rethinking technological change in smallholder agriculture. *Outlook on Agriculture, 48*(3), 169–180. https://doi.org/10.1177/0030727019864978

Goldman, A., & Smith, J. (1995). Agricultural transformations in India and Northern Nigeria: Exploring the nature of green revolutions. *World Development, 23*(2), 243–263. https://doi.org/10.1016/0305-750X(94)00115-F

Gregoire, A. (2002). The mental health of farmers. *Occupational Medicine, 52*, 471–476.

Griffifin, K. (1979). *The political economy of agrarian change: An essay on the green revolution* (2nd ed.). Macmillan Press.

Grzywacz, J. G., Quandt, S. A., Chen, H., Isom, S., Kiang, L., Vallejos, Q., & Arcury, T. A. (2010). Depressive symptoms among Latino farmworkers across the agricultural season: Structural and situational influences. *Cultural Diversity & Ethnic Minority Psychology, 16*, 335–343.

Hanklang, S., Kaewboonchoo, O., Morioka, I., & Plernpit, S.-A. (2016). Gender difffferences in depression symptoms among rice farmers in Thailand. *Asia-Pacific Journal of Public Health, 28*, 83–93.

Hansson, G. A., Asterland, P., Holmer, N. G., & Skerfving, S. (2001). Validity and reliability of triaxial accelerometers for inclinometry in posture analysis. *Medical & Biological Engineering & Computing, 39*(4), 405–413. https://doi.org/10.1007/bf02345361

Hansson, G. A., Balogh, I., Ohlsson, K., Granqvist, L., Nordander, C., Arvidsson, I., Akesson, I., Unge, J., Rittner, R., Stromberg, U., & Skerfving, S. (2010). Physical workload in various types of work: Part II. Neck, shoulder and upper arm. *International Journal of Industrial Ergonomics, 40*(3), 267–281. https://doi.org/10.1016/j.ergon.2009.11.002

Harrison, V., & Mackenzie Ross, S. (2016). Anxiety and depression following cumulative low-level exposure to organophosphate pesticides. *Environmental Research, 151*, 528–536.

Hart, S. G. (2006). NASA-task load index (NASA-TLX); 20 years later. *NASA-Ames Research Center, 50*, 904–908.

Hayami, Y., & Ruttan, V. W. (1970). Factor prices and technical change in agricultural development: The United States and Japan, 1880–1960. *Journal of Political Economy, 78*(5), 1115–1141.

Hazell P (2009) *The Asian Green revolution*. IFPRI discussion paper 00911, Washington.

Hazell, P., & Ramasamy, C. (1991). *The green revolution reconsidered: The impact of high-yielding rice varieties in South India*. Johns Hopkins University Press.

Hedlund, D., & Berkowitz, A. (1979). The incidence of socialpsychologicl stress in farm families. *International Journal of Sociology of the Family, 9*, 233–243.

Herderschee, J., Kaiser, K.-A., & Mukoko, D. S. (2012). *Resilience of an African giant: Boosting growth and development in the Democratic Republic of Congo*. The World Bank.

Hiott, A. E., Grzywacz, J. G., Davis, S. W., Quandt, S. A., & Arcury, T. A. (2008). Migrant farmworker stress: Mental health implications. *The Journal of Rural Health, 24*, 32–39.

Hong, Z.-R., Hong, S.-Y., Han, M.-J., Lee, H.-S., Gil, H.-O., Yang, J.-O., Lee, E.-Y., & Hong, S.-Y. (2008). Clinical observation of 12 farmers who believe themselves to have suffered from chronic pesticide intoxication. *The Korean Journal of Internal Medicine, 23*, 1–4.

Hong, S.-Y., Jeong, D.-S., Gil, H.-W., Yang, J.-O., Lee, E.-Y., & Hong, S.-Y. (2009). The estimation of pesticide exposure in depression scores: In case of Korean orchard farmers. *Journal of Pest Science, 82*, 261–265.

Hossain, D., Eley, R., Coutts, J., & Gorman, D. (2008). Mental health of farmers in Southern Queensland: Issues and support. *The Australian Journal of Rural Health, 16*, 343–348.

Hovey, J. D., & Magaña, C. G. (2002a). Exploring the mental health of Mexican migrant farm workers in the Midwest: Psychosocial predictors of psychological distress and suggestions for prevention and treatment. *The Journal of Psychology, 136*, 493–513.

Hovey, J. D., & Magaña, C. G. (2002b). Psychosocial predictors of anxiety among immigrant Mexican migrant farmworkers: Implications for prevention and treatment. *Cultural Diversity & Ethnic Minority Psychology, 8*, 274–289.

Irz, X., Lin, L., Thirtle, C., & Wiggins, S. (2001). Agricultural productivity growth and poverty alleviation. *Development and Policy Review, 19*(4), 449–466. https://doi.org/10.1111/1467-7679.00144

Jamal, G. A., Hansen, S., Pilkington, A., Buchanan, D., Gillham, R. A., Abdel-Azis, M., Julu, P. O. O., Al-Rawas, S. F., Hurley, F., & Ballantyne, J. P. (2002). A clinical neurological, neurophysiological, and neuropsychological study of sheep farmers and dippers exposed to organophosphate pesticides. *Occupational and Environmental Medicine, 59*, 434.

Johannsen, D. L., Calabro, M. A., Stewart, J., Franke, W., Rood, J. C., & Welk, G. J. (2010). Accuracy of armband monitors for measuring daily energy expenditure in healthy adults. *Medicine and Science in Sports and Exercise, 42*, 2134–2140. https://doi.org/10.1249/MSS.0b013e3181e0b3ff

Kallioniemi, M. K., Simola, A. J. K., Kymäläinen, H.-R., Vesala, H. T., & Louhelainen, J. K. (2009). Mental symptoms among Finnish farm entrepreneurs. *Annals of Agricultural and Environmental Medicine, 16*, 159–168.

Kallioniemi, M. K., Simola, A., Kaseva, J., & Kymäläinen, H.-R. (2016). Stress and burnout among finnish dairy farmers. *Journal of Agromedicine, 21*, 259–268.

Karpansalo, M., Manninen, P., Lakka, T. A., Kauhanen, J., Rauramaa, R., & Salonen, J. T. (2002). Physical workload and risk of early retirement: Prospective population-based study among middle-aged men. *Journal of Occupational and Environmental Medicine, 44*, 930–939.

Kearney, G. D., Rafffferty, A. P., Hendricks, L. R., Allen, D. L., & Tutor-Marcom, R. (2014). A cross-sectional study of stressors among farmers in eastern North Carolina. *North Carolina Medical Journal, 75*, 384–392.

Kim, J., Ko, Y., & Lee, W. J. (2013). Depressive symptoms and severity of acute occupational pesticide poisoning among male farmers. *Occupational and Environmental Medicine, 70*, 303.

Kunde, L., Kõlves, K., Kelly, B., Reddy, P., & De Leo, D. (2017). Pathways to suicide in Australian farmers: A life chart analysis. *International Journal of Environmental Research and Public Health, 14*, 352.

Kureshi, J. S., & Somsundaram, K. (2018). Assessment of occupational stress among farmers in Aurangabad district, Maharashtra. *International Journal of Community Medicine and Public Health, 5*, 1434–1440.

Langley, R. L., & Morrow, W. E. M. (2010). Livestock handling-minimizing worker injuries. *Journal of Agromedicine, 15*(3), 226–235. https://doi.org/10.1080/1059924x.2010.486327

Lazarus, R. S. (1999). *Stress and emotion: A new synthesis*. Springer.

Linn, J. G., & Husaini, B. A. (1987). Determinants of psychological depression and coping behaviors of Tennessee farm residents. *Journal of Community Psychology, 15*, 503–512.

Liu, B.-P., Qin, P., Liu, Y.-Y., Yuan, L., Gu, L.-X., & Jia, C.-X. (2018). Mental disorders and suicide attempt in rural China. *Psychiatry Research, 261*, 190–196.

Logstein, B. (2016). Farm-related concerns and mental health status among Norwegian farmers. *Journal of Agromedicine, 21*, 316–326.

Lunner Kolstrup, C., Kallioniemi, M., Lundqvist, P., Kymäläinen, H.-R., Stallones, L., & Brumby, S. (2013). International perspectives on psychosocial working conditions, mental health, and stress of dairy farm operators. *Journal of Agromedicine, 18*, 244–255.

Malekirad, A. A., Faghih, M., Mirabdollahi, M., Kiani, M., Fathi, A., & Abdollahi, M. (2013). Neurocognitive, mental health, and glucose disorders in farmers exposed to Organophosphorus pesticides (Report). *Archives Hygiene Toksikol, 64*, 1–8.

McIntyre, B. D., Herren, H. R., Wakhungu, J., et al. (2009). *Agriculture at a Crossroads: IAASTD Global Report*. Island Press.

McShane, C. J., Quirk, F., & Swinbourne, A. (2016). Development and validation of a work stressor scale for Australian farming families. *The Australian Journal of Rural Health, 24*, 238–245.

Mendis, I. U., & Udomsade, J. (2005). Factors affecting adoption of recommended crop management practices in Paddy cultivation in Kaluytara District. *Srilanka Kasetsart Journal (Social Science), 26*, 91–102.

Miller, R., & Brown, W. (2004). Steps and sitting in a working population. *International Journal of Behavioral Medicine, 11*, 219–224.

Mora, D. C., Quandt, S. A., Chen, H., & Arcury, T. A. (2016). Associations of poor housing with mental health among North Carolina Latino migrant farmworkers. *Journal of Agromedicine, 21*, 327–334.

Morris, M., Kelly, V., Kopicki, R. J., & Byerlee, D. (2007). *Fertilizer use in African agriculture: Lessons learned and good practice guidelines*. The World Bank.

Muñoz-Quezada, M. T., Lucero, B., Iglesias, V., Levy, K., Muñoz, M. P., Achú, E., Cornejo, C., Concha, C., Brito, A. M., & Villalobos, M. (2017). Exposure to organophosphate (OP) pesticides and health conditions in agricultural and non-agricultural workers from Maule. *Chile International Journal of Environment and Health Research, 27*, 82–93.

Ngai, R., & Pissarides, C. (2007). Structural change in a multi-sector model of growth. *The American Economic Review, 97*(1), 429–443.

O'Connor, K., Stoecklin-Marois, M., & Schenker, M. B. (2015). Examining nervios among immigrant male farmworkers in the MICASA study: Sociodemographics, housing conditions and psychosocial factors. *Journal of Immigrant and Minority Health, 17*, 198–207.

OECD. (2012). *Agricultural policies for poverty reduction*. OECD Publishing. https://doi.org/10.1787/9789264112902-en

Olowogbon, T. S., Yoder, A. M., Fakayode, S. B., & Falola, A. O. (2019). Agricultural stressors: Identification, causes and perceived effects among Nigerian crop farmers. *Journal of Agromedicine, 24*, 46–55.

Onwuameze, O. E., Paradiso, S., Peek-Asa, C., Donham, K. J., & Rautiainen, R. H. (2013). Modifiable risk factors for depressed mood among farmers. *Annals of Clinical Psychiatry, 25*, 83–90.

Page, A. N., & Fragar, L. J. (2002). Suicide in Australian farming. *Australian and New Zealand Journal of Psychiatry, 36*, 81–85.

Pattnaik, S., & Mohapatra, D. (2019). Modern Technology in Agriculture and Perception of the farmers: A case study in Odisha. *International Journal of Applied Social Science, 6*(1), 49–55.

Peel, D., Berry, H. L., & Schirmer, J. (2015). Perceived profitability and well-being in Australian dryland farmers and irrigators. *The Australian Journal of Rural Health, 23*, 207–214.

Perceval, M., Kõlves, K., Reddy, P., & De Leo, D. (2017). Farmer suicides: A qualitative study from Australia. *Occupational Medicine, 67*, 383–388.

Povey, A. C., McNamee, R., Alhamwi, H., Stocks, S. J., Watkins, G., Burns, A., & Agius, R. (2014). Pesticide exposure and screen-positive neuropsychiatric disease in British sheep farmers. *Environmental Research, 135*, 262–270.

Prax V (2010) *American family farmers feed 155 people each-2% Americans farm*. Retrieved from http://suite101.com/article/american/family-farmers-feeds-155-people-each-2-americans-farm-a231011

Price, L., & Evans, N. (2009). From stress to distress: Conceptualizing the British family farming patriarchal way of life. *Journal of Rural Studies, 25*, 1–11.

Pulgar, C., Trejo, G., Suerken, C., Ip, E., Arcury, T., & Quandt, S. (2016). Economic hardship and depression among women in Latino farmworker families. *Journal of Immigrant and Minority Health, 18*, 497–504.

Quandt, S. A., Chen, H., Grzywacz, J. G., Vallejos, Q. M., Galvan, L., & Arcury, T. A. (2010). Cholinesterase depression and its association with pesticide exposure across the agricultural season among Latino farmworkers in North Carolina. *Environmental Health Perspectives, 118*, 635–639.

Raine, G. (1999). Causes and effects of stress on farmers: A qualitative study. *Health Education Journal, 58*, 259–270.

Ramos, A., Su, D., Lander, L., & Rivera, R. (2015). Stress factors contributing to depression among Latino migrant farmworkers in Nebraska. *Journal of Immigrant and Minority Health, 17*, 1627–1634.

Rocha, L. P., Borges, A. M., Bonow, C. A., de Almeida, M. C. V., Piexak, D. R., & Cezar-Vaz, M. R. (2016). Rural workload: Study performed in two rural environments in the state of Rio Grande do Sul, Brazil. *Ciência Rural, 46*(5), 832–838.

Rogers, E. M. (2003). *Diffusion of innovations* (5th ed.). Free Press.

Roy, P., Tremblay, G., Oliffffe, J. L., Jbilou, J., & Robertson, S. (2013). Male farmers with mental health disorders: A scoping review. *The Australian Journal of Rural Health, 21*, 3–7.

Sahu, N. C., & Mishra, D. (2014). Economic impact of climate change on agriculture sector of Coastal Odisha. *APCBEE Procedia, 10*, 421–425.

Saka, J. O., Okoruwa, V. O., Lawal, B. O., & Ajijola, S. (2005). Adoption of improved Rice varieties among small holder farmers in South-Western Nigeria. *World Journal of Agriculture Science, 1*(1), 42–49.

SECCO IAO, et al. (2011). Workloads of external materiality in the nursing team of a teaching hospital in Paraná, Brazil. *Cienciay Enfermería, 17*, 69–81.

Siegel, M., Starks, S., Sanderson, W., Kamel, F., Hoppin, J., & Gerr, F. (2017). Organic solvent exposure and depressive symptoms among licensed pesticide applicators in the agricultural health study. *International Archives of Occupational and Environmental Health, 90*, 849–857.

Simkin, S., Hawton, K., Fagg, J., & Malmberg, A. (1998). Stress in farmers: A survey of farmers in England and Wales. *Occupational and Environmental Medicine, 55*, 729.

Simmons, J. (2011). Technology's role in the 21st century: Making safe, affordable and abundant food a global reality. Elanco animal. *Health*, 1–12.

Simsek, Z., Ersin, F., & Kirmizitoprak, E. (2015). Development of the seasonal migrant agricultural worker stress scale in Sanliurfa. *Southeast Turkey Journal of Agromedicine, 21*, 56–60.

Singh, S., Hanna, E. G., & Kjellstrom, T. (2013). Working in Australia's heat: Health promotion concerns for health and productivity. *Health Promotion International, 30*, 239–250.

Smale, M. (1995). "Maize is life": Malawi's delayed Green revolution. *World Development, 23*(5), 819–831. https://doi.org/10.1016/0305-750X(95)00013-3

Soer, R., van der Schans, C. P., Groothoff, J. W., Geertzen, J. H., & Reneman, M. F. (2008). Towards consensus in operational definitions in functional capacity evaluation: A Delphi survey. *Journal of Occupational Rehabilitation, 18*, 389–400. https://doi.org/10.1007/s10926-008-9155-y

Soer, R., van der Schans, C. P., Geertzen, J. H., Groothoff, J. W., Brouwer, S., Dijkstra, P. U., et al. (2009). Normative values for a functional capacity evaluation. *Archives of Physical Medicine and Rehabilitation, 90*, 1785–1794. https://doi.org/10.1016/j.apmr.2009.05.008

Staniford, A. K., Dollard, M. F., & Guerin, B. (2009). Stress and help-seeking for drought-stricken citrus growers in the Riverland of South Australia. *The Australian Journal of Rural Health, 17*, 147–154.

Steele, R., & Mummery, K. (2003). Occupational physical activity across occupational categories. *Journal of Science and Medicine in Sport, 6*, 398–407.

Steenland, K., Jenkins, B., Ames, R. G., O'Malley, M., Chrislip, D., & Russo, J. (1994). Chronic neurological sequelae to organophosphate pesticide poisoning. *American Journal of Public Health, 84*, 731–736.

Suten Geofrey, M., Ezra Jonathan, M., Aiwerasia Vera, N., & Simon, M. (2018). Health symptoms associated with pesticides exposure among flower and onion pesticide applicators in Arusha region. *Annals of Global Health, 84*, 369–379.

Temple, L., Bienabe, E., Barret, D., et al. (2016). Methods for assessing the impact of research on innovation and development in the agriculture and food sectors. *African Journal of Science, Technology, Innovation and Development, 8*, 399–410.

Terrazas, S. R., & McCormick, A. (2018). Coping strategies that mitigate against symptoms of depression among Latino farmworkers. *Hispanic Journal of Behavioral Sciences, 40*, 57–72.

Thirtle, C., Lin, L., & Piesse, J. (2003). The impact of research-led agricultural productivity growth on poverty reduction in Africa. *Asia and Latin America World Develpment, 31*(12), 1959–1976. https://doi.org/10.1016/j.worlddev.2003.07.001

Thornton, P. K., Schuetz, T., Forch, W., et al. (2017). Responding to global change: A theory of change approach to making agricultural research for development outcome-based. *Agricultural Systems, 152*, 145–153.

Thorp, A. A., Healy, G. N., Winkler, E., Clark, B. K., Gardiner, P. A., Owen, N., et al. (2012). Prolonged sedentary time and physical activity in workplace and non-work contexts: A cross-sectional study of office, customer service and call Centre employees. *International Journal of Behavioral Nutrition and Physical Activity, 9*, 128. https://doi.org/10.1186/1479-5868-9-128

Tizale CY (2007) *The dynamics of soil degradation and incentives for optimal management in the central highlands of Ethiopia*. PhD thesis, Department of Agricultural Economics, Extension and Rural Development, Faculty of Natural and Agricultural Sciences, University of Pretoria.

Truchot, D., & Andela, M. (2018). Burnout and hopelessness among farmers: The farmers stressors inventory. *Social Psychiatry and Psychiatric Epidemiology, 5*, 1–9.

Tudor-Locke, C., Craig, C. L., Thyfault, J. P., & Spence, J. C. (2013). A step-defined sedentary lifestyle index: <5000 steps/day. *Applied Physiology, Nutrition, and Metabolism, 38*, 100–114. https://doi.org/10.1139/apnm-2012-0235

Tuomisto, H. L., Scheelbeek, P. F. D., Chalabi, Z., Green, R., Smith, R. D., Haines, A., & Dangour, A. D. (2017). Effects of environmental change on agriculture, nutrition and health: A framework with a focus on fruits and vegetables. *Wellcome Open Research, 2*, 21. https://doi.org/10.12688/wellcomeopenres.11190.2

Valluru, C. T., Dekker, S., & Rae, A. (2017). How and why do subcontractors experience different safety on high-risk work sites? *Cognition, Technology & Work, 19*, 785–794. https://doi.org/10.1007/s10111-017-0435-1

Vega, W., Warheit, G., & Palacio, R. (1985). Psychiatric symptomatology among Mexican American farmworkers. *Social Science & Medicine, 20*, 39–45.

Walker, L. S., & Walker, J. L. (1987). Stressors and symptoms predictive of distress in farmers. *Family Relations, 36*, 374–378.

Walker, J. L., Walker, L. S., & MacLennan, P. M. (1986). An informal look at farm stress. *Psychological Reports, 59*, 427–430.

Weisskopf, M. G., Moisan, F., Tzourio, C., Rathouz, P. J., & Elbaz, A. (2013). Pesticide exposure and depression among agricultural Workers in France. *American Journal of Epidemiology, 178*, 1051–1058.

Wesseling, C., van Wendel de Joode, B., Keifer, M., London, L., Mergler, D., & Stallones, L. (2010). Symptoms of psychological distress and suicidal ideation among banana workers with a history of poisoning by organophosphate or n-methyl carbamate pesticides. *Occupational and Environmental Medicine, 67*, 778–781.

Wheeler, S. A., Zuo, A., & Loch, A. (2018). Water torture: Unravelling the psychological distress of irrigators in Australia. *Journal of Rural Studies, 62*, 183–194.

Wiatrowski, W. J., & Janocha, J. A. (2014). *Comparing fatal work injuries in the United States and the European Union. Monthly Labor Review*. U.S. Bureau of Labor Statistics. https://doi.org/10.21916/mlr.2014.23

Williams, R. T. (1996). The on-going farm crisis: Extension leadership in rural communities. *Journal of Extension, 34*, 1–3.

Williams, R. (2001). The ongoing farm crisis: Health, mental health and safety issues in Wisconsin. *Rural Mental Health, 26*, 15–17.

World Health Organization. (2007a). *Depression and other common mental disorders: Global Health estimates*. World Health Organization.

World Health Organization. (2007b). *Promoting mental health: Concepts, emerging evidence, practice: Summary report*. World Health Organization.

Zhang, X., Wu, M., Yao, H., Yang, Y., Cui, M., Tu, Z., Stallones, L., & Xiang, H. (2016). Pesticide poisoning and neurobehavioral function among farm workers in Jiangsu, People's Republic of China. *Cortex, 74*, 396–404.

Chapter 3
Features of 'Internet of Things' to Resolve Agricultural Problems

3.1 Introduction

Agricultural part is a vital however dismissed division throughout the world. A man inclines increasingly more towards mechanized and pleasant life. So human improvement is progressively pulled in towards the mechanical division. Agriculture is an old customary business, yet simultaneously it is a non-beneficial business-fragment. Work related safety is a significant issue of conversation for cultivating experts. The systems for working in field in extraordinary climates, in-contact with synthetic compounds like pesticides & composts, exposures to soils, dusts and animals, contamination due to bacteria, injuries due to hand-tools and musculoskeletal disorders are the most significant issues looked by every single agrarian specialist. In the execution of various agricultural operations starting from preparing of lands to post-harvesting operations, the agricultural workers play a crucial role. The farmers are found to use different types of farm-related tools and equipment for performing different crop-growing activities. However, the efficient use of these tools and appliances requires a good knowledge. Proper design of equipment is more essential for increasing work-efficiency and safety-levels of farmers. In the farms of most of the developing countries, the hand-tools are commonly used and normally, the farmers are exposed to various hazards due to musculoskeletal risk-factors, machinery, organic gases & dusts, pesticides, and others involved in agricultural activities.

Farmers working on farms understand that the tools and equipment they use can be dangerous, if not properly used. Unfortunately, many farmers are injured due to carelessness, lack-of-knowledge about the use of tools, and in some cases the injuries or accidents are beyond their controls which are caused because of defective as well as in-effective design of farm-tools and equipment. Because of exposing to agro-chemicals, psycho-social stresses, poor-ergonomics and poor-physical working-conditions, the farmers were attributed to various hazards (White & Cessna, 1989). Although a little extent of the workforce is utilized in cultivating, but

S. Satapathy et al., *Innovation in Agriculture with IoT and AI*,
Innovation, Technology, and Knowledge Management,
https://doi.org/10.1007/978-3-030-88828-2_3

a greater percentage of all fatal accidents in workplaces occur in agricultural sectors. Larger ranges of health and safety issues apparent to workplace hazards, miscellaneous health and safety programs, and policies & practices, are mostly dealt with occupational health and safety (OHS). The occupational health surveillance programs in agriculture help to recognize early indications of work-related infirmity among farmers, such that appropriate anticipatory measures to improve health and safety of workers can be taken up (Tabibi et al., 2017). As OHS is regarded to be most essential from economic, moral, and legitimate purpose of perspectives, thus taking commencements becomes the united duty of working groups, employers, in addition to the government. Additionally, in modern segments as well as in agribusiness, the OHS has turned into a note-worthy for overall concerns.

Different undesirable consequences from a variety of hazards are usually associated in agriculture such as infection by pathogens, musculoskeletal disorders, injuries from exposures, poisoning, physiological-disorders, and respiratory-infections. Moreover, agriculture has been ranked among the top three major perilous occupational groups in terms of work-related illnesses, injuries, and fatalities, by the "Centers for Disease Control and Prevention" (CDC, 2013). Moreover, for nourishment creation, the farmers need to work in the out-doors on a daily-basis and all the more frequently perform rehashed relentless physical exercises for extended time periods in awkward and un-comfortable working postures (Pyykkönen & Aherin, 2012). Heavy physical-workloads as well as ergonomically poor working-conditions have been regarded as the pre-dominant factors for musculoskeletal disorders (MSDs) and work-related injuries (Giuffrida et al., 2001). The absence of ergonomic contemplation in the design of hand-tools and equipment, lack of skills, absence of the information of wellbeing measures and negligence of farmers and un-favorable conditions may lead to agricultural injuries and accidents (Mishra & Satapathy, 2018a).

Moreover, the objects are connected with sensors or controlled remotely through "Internet of Things (IoT)" across existing networks that improves efficiency, results in more accuracy and economic benefits as well as reduction of human interventions. The applications of IoT can be evolved between the peoples, between people and things in addition to between things and things. Accuracy in agriculture can be obtained through IoT that enables items to be controlled remotely over the framework of existing systems thus bring about improved effectiveness and financial advantages. Therefore, this study aimed at discussing and assessing the problems associated with ergonomic risks as well as MSDs among the farmers along with possible measures through the application of IoT in agricultural machinery for the benefits of developing as well as under-developed countries.

3.2 Ergonomic and Posture Analysis in Farming

The augmentation of the connection of the farmers with their corresponding places of work is highly desirable. Moreover, the enhancement of productivity, working comforts as well as efficiency of farm workers with due emphasis on their well-being and safety can be achieved by proper consideration of various ergonomic aspects in the design and use of the agricultural tools and equipment. An investigation was made on different harvesters' vibrations of at both idle as well as full-load conditions. It was observed that in 10% of the exposed population, the "traumatic-vasospastic-disease" was appeared after 0.7 to 7.1 years for the left-hands, whereas it appeared after 1.0 to 4.7 years in the right-hands of the operators using continuously (Cakmak et al., 2011). Dooley (2012) has detailed that an ergonomic mismatch prompts numerous item dissensions and disappointments. An effective item in the present agricultural segment requires great comprehension of ergonomics. Dominant part of the provincial population was actively associated with farming sectors. At the same time the farmers having little possessions were found to do a large portion of the farming exercises by utilizing manual techniques, which resulted in huge wastage of human work, and thus bringing down the yield per-capita workforce. Villarejo and Baron (1999) have revealed that the farm machines and tools associated vibrations make the farmers to more often suffer from back pains, and pains in their shoulders, arms, and hands, respectively. Moreover, it has been accounted for that 33% of the injuries that makes farmers to avoid cultivate works were sprains and strains, while a quarter was due to back injuries. From a study focused on the women workforce engaged in tea industry in Assam, the musculoskeletal issues were reported at different body parts because of repetitive and awkward postures by utilizing the existing basket. Subsequently, with the modification of the basket in light of anthropometric information, a noteworthy decrease in inconvenience at various body parts were reported (Bhattacharyya & Chakrabarti, 2012). Jatinder and Rana (2012) have made an investigation on farm women during the weeding activities on hill farms by using traditional tools, as well as three modified and improved weeding tools according to anthropometric information. Further the improved tools were provided with convenient handles and made light in weight with sharp inward edges. The relative appraisal between the utilization of conventional and enhanced apparatuses demonstrated better outcomes as far as pulse esteems.

The agricultural sectors not only supplies food and nourishment,but also help in utilizing various labourer (Naeini et al., 2014). Agri-business has been perceived as the most perilous industry with a high rate of musculoskeletal disorders (MSDs) in the country. The utilization of ergonomic methodologies in agricultural appliances configuration was reported to be additionally restricted to a couple (Vyavahare & Kallurkar, 2012). As a result, the misuse in the accessible-assets in addition to the innovations at ideal-level by proper utilization of best ergonomic-practices in the farming sectors has been an exceptional requirement to improve the efficiency. In a study, the issues of the utilization of power-tillers for various agricultural activities by the farmers was considered, and it was seen that abnormal-state of vibrations from the power-tillers was transmitted from the handles to hands, arms and shoulders, and thus by making early-exhaustion of the operators. However, for decreasing the magnitude of vibrations, three materials were utilized for mediation improvement like polyurethanes, rubbers, and blend of polyurethanes& rubbers. It was additionally discovered that the greatest decrease in vibrations were accomplished with the use of elastic when contrasted with polyurethanes, and blend of polyurethanes as well as rubbers (Chaturvedi et al., 2012). Nunes (2009) has distinguished three sorts of hazard-factors which can be considered in the survey of the work-related MSDs as "Physical factors e.g. continuous and in-comfortable postures, repetitive-movements, forceful-activities, vibrations of hand and arms, vibrations of entire-body, mechanical-pressure, and cold"; "Psychosocial factors e.g. work-rate, work or rest cycles, task-demands, social-support and job-uncertainty"; "Individual factors e.g. age, sexual-orientation, proficient-exercises, sport-exercises, residential-exercises, recreational-exercises, liquor/tobacco utilization, and past work-related MSDs".

Distinctive ergonomic assessment tools are utilized to assess the ergonomic danger of work or task. For example, the "Quick Exposure Check (QEC)", "Rapid Upper Limb Assessment (RULA)", and the "Rapid Entire Body Assessment (REBA)" are more utilized ergonomic risk assessment tools that measure the ergonomic dangers/risks of both lower and upper parts of the human body. The RULA is utilized to delineate the acceptability of different tasks & postures, and recommends whether the assignment or posture is worthy or should be explored further or should be changed before long in light of RULA score values. Considering various variables and user data, RULA makes the postures examination in light of the parameters such as distance, weight and frequency. RULA was developed to show a quick measure of work-related musculoskeletal disorders (WRMSDs) in which the worker performs work with constant postures, and the work related upper limb issue were for the most part reported. RULA was broadly used to evaluate the postures, forces and movements related with different tasks and the outcomes were shown as risk scores going from one to seven, where more prominent levels of clear risk were demonstrated by higher score values (McAtamney & Corlett, 1993). In order to optimize different human-postures, the RULA analysis technique was discussed for accomplishing better designs, satisfactory items and workplaces (Ren & Xiao, 2009; Sanjog et al., 2012). Dube et al. (2018) have used RULA tool for an ergonomic evaluation of working postures of Indian farmers. The different processes considered were seeding, fertilization, and weeding, and moreover, based on the obtained RULA scores, different activity-levels were recommended accordingly for the working methods. For the evaluation and analysis of the postures in working and the workloads, the method by the "Finish steel industry" such as the "Ovako Working Posture Analysing System (OWAS)" has been developed (Karhu et al., 1977, 1981), and this method was found its successful application in the ergonomic field (Brandl et al., 2017a; Hellig et al., 2018). In six tea factories, forty-eight workers were considered to analyze their postures by the use of methods such as REBA and OWAS. Moreover, 75% of different postures of the workers were reported of needing corrective measures. Whereas 34% of various postures taken by the workers needed corrective measures. 34% of different postures were reported of at high and very high-risk levels on the basis of OWAS result (Soheili-Fard et al., 2017).

3.2.1 *Posture Analysis Using Ergonomic Tool*

In order to assess the working postures of farmers for different farming activities in three developing countries like "India, Nepal and Bangladesh", the RULA tool was used. Further recommendations of action-levels were made by considering the obtained RULA scores. Initially by the use of a digital-camera, the still-photographs of farmers were taken by considering the common-postures taken during different farming activities. Both traditional as well as machine-based techniques in cultivation practices were taken into consideration while taking the photographs. Then, the "ErgoFellow-3.0 software" was utilized for performing the postural analysis and the RULA scores were obtained accordingly.

The different farming activities that were considered (Fig. 3.1), included: "Preparing land by the use of bullock-drawn ploughs; Tractor-implements for ploughing; Fertilizers' application; Pesticides-spraying; Manual weed-pulling; and Winnowing"; respectively.

It was observed that in order to prepare the land, the farmers were required to attach the ploughs to the bullocks followed by guiding the bullock-driven ploughs in the agricultural field (Figure 3.1A). During this activity, the considered farmer was required to carry a static-load lesser than 2 kg in operating the plough with intermittently bending-forward, and with other postures like bending of knees,

Fig. 3.1 Different farming activities

forward-bending of neck at $0–15^0$, bending as well as twisting of elbows as well as wrists of both-hands. The corresponding RULA score obtained was 7 based on these postures. However, when the tractors were attached with the ploughs as shown in Figure 3.1B, the tractor-operator was observed of bending knees, bending as well as twisting of the wrists of both-hands at15^0+, and bending neck at $10^0–20^0$. Based on these postures, the corresponding RULA score was 4.

During manual fertilizers application activities, it was observed that the farmers were required to move around the farm by sprinkling the fertilizers by the use of hands (Figure 3.1C). The observed postures were: carrying of a static-load of 2-10 kg, lower-arm at $0–60^0$, forward-bending of neck at $0–20^0$, and corresponding RULA score of 5 was obtained. While performing the pesticides-spraying activities with manual-techniques by holding as well as carrying the pesticide-sprayers (Figure 3.1D), it was observed that the farmers were required of lowering the arms at $0–60^0$, forward-bending of neck at $0–20^0$, twisting the elbows as well as wrists of both-hands, carrying a static-load lesser than 10 kg, and a flexion of upper as well as forearms, respectively. The obtained RULA score was 6 based on these postures.

When performing the manual weed-pulling activities by the utilization of traditional sickles or knifes as shown in Figure 3.1E, the farmers were observed of bending-forward at $20^0–60^0$, working across midline of body, bending of knees, bending of neck, bending as well as twisting of the elbows as well as wrists of both-hands, and having flexion of upper as well as forearms with little-loads i.e. lesser than 2 kg in hands. Moreover, a RULA score of 7was obtained based on these postures. Moreover, while performing the winnowing activities, it was observed that the farmers were intermittently bending forward, bending the knees, bending neck at 20^0+, bending as well as twisting the elbows in addition to wrists of both-hands, and having a flexion of upper as well as forearms with little-loads in hands as shown in Figure 3.1F. In view of these postures, a corresponding RULA score of 5 was obtained.

Table 3.1 Obtained RULA scores' summary

Activities		RULA score	Proposed-actions
Agricultural land's preparations	(a) Preparing land by the use of bullock-drawn ploughs	7	Investigation and changes are required immediately
	(b) Tractor-implements for ploughing	4	Further investigation is needed and changes may also be required
Fertilizers' application		5	Investigation and changes are required soon
Pesticides-spraying		6	Investigation and changes are required soon
Manual weed-pulling		7	Investigation and changes are required immediately
Winnowing		5	Investigation and changes are required soon

On the basis of the obtained RULA scores in various agricultural activities considered, different action-levels were proposed accordingly as illustrated in Table 3.1.

In spite of rapid growth in technologies, the traditional tools and techniques were mostly found as the only-options in order to accomplish the farming activities for rural-farmers of most of the developing as well as under-developed countries. Hence for the benefits in farming-sectors, respective competent authorities must have a high-vision for the farmers to provide them with the latest tools and techniques, providing training regarding the operation and safety measures, providing adequate medical-facilities, and regular-inspections of respective activities. However, the objectives of the agricultural sectors would not be achieved without giving proper-consideration to ergonomic-aspects. Therefore, proper as well as effective applications of ergonomics in the farming-sectors can help in achieving a balance by connecting the characteristics of farmers with the demands of tasks. Thus, through appropriate ergonomic considerations in the design of farm-related tools and equipment, the safety, job-satisfactions in workplaces, productivity, and physical as well as mental wellbeing of farmers can be enhanced to a greater extent.

3.3 MSDs Problems in Agri-Farming

Although, different studies were made considering the injuries in agricultural sectors worldwide, but the studies on agricultural injuries due to hand tools and equipment in particular are limited to a few. For instance, MSDs were reported of occurring due to the perceived discomfort in using hand tools for longer term (Hamberg-van Reenen et al., 2008). Patel et al. (2010) have stated that because of non-availability of compiled information on agricultural injuries in nationwide, currently there was a lack of study on agricultural injuries from developing countries. The number of injuries was reported as 214 for male and 109 for female farmers during the period 2006–2010, in "West Bengal (India)". The hand tools related injuries were 64.7%, farm machinery related were 29.1% and others as 6.2%. However, spade and sickle were reported as the most frequently involved tools for hand injuries. It was further observed that the male farmers were much more affected than the female farmers, and the accident rate among male farmers was 8.99 per thousand workers per year, while for female farmers was 7.89 per thousand workers per year, respectively (Das, 2014). Jain et al. (2016) have investigated the practice of QFD for developing the hand-tools in the agriculture and in other areas by using concepts of ergonomics principles. Mishra and Satapathy (2018b) have conducted a survey about the agricultural injuries of farmers in India, during the period from 2014 to 2017. It was reported that maximum number of male and female victim farmers were in the age group of 31 to 45, and also 43 farmers out of 45 farmers were the victims of agricultural injuries by the use of the agricultural hand tools such as knifes, pick-axe, weeding-fork, spades, plain-edge sickles, serrated sickles, and shovels, respectively. It has been revealed for the sustainable-development that the need-based agricultural tools and equipment design based on the ergonomic aspects would have a direct multiplier-effect on the socio-economic farmers' transformation (Patel & David, 2016). Parvez and Shahriar (2018) have studied the main causes of agricultural injuries with the farmers in "Panchagarh district of Bangladesh". It was reported of about 67% injuries due to hand-tools and remaining 33% due to machinery as well as other sources.

The lesser technically advanced sector workers (i.e., handicrafts, agriculture, constructions, and so on), especially involved in consistent postures and repetitive manual tasks, were reported of suffering of various work-related problems because of the factors such as individual, work characteristics and tool-related factors (Osborne et al., 2012; Jain et al., 2018). Different authors and researchers have reported of MSDs of farmers in the agricultural sectors to be highly affected by the risk-factors evolved in the respective workplaces. Farmers were reported of suffering from various work-related problems in the agricultural sectors which is a challenging occupation and MSDs were revealed as the common problem (Fathallah, 2010). Osborne et al. (2012) have reported of lower back-pain to be the most general MSDs among the farmers that were followed by upper as well as lower extremity muscle disorders. It was further revealed of a greater prevalence of MSDs for farmers as compared to the non-farming individuals. As the farmers usually adopt postures in which trunk & head move forward with slight inclinations during the manual harvesting for better holding as well as cutting of crops, thus they get exposed to the risk of MSDs development. It was reported that the operations such as weeding and harvesting of crops result in excessive physical strain which may further contribute in developing MSDs in the farmers (Das & Gangopadhyay, 2011; Das, 2015). Kumar et al. (2008) have considered forty-two numbers of villages of four-districts in Arunachal-Pradesh, India and after making an investigation about the agricultural accident for six-years, i.e., considering a period of '2000 to 2005', they have reported of the occurrence of 40% farm-implement related-injury with the accident-rate to be 6.39 per thousand-workers per year. The hand has been reported as the most commonly injured-part compared to the lower-limb and foot in a study made on the agricultural injuries of 41 patients for three-year period (Copuroglu et al., 2012). Wibowo and Soni (2016) have considered the farmers of "East-Java and Indonesia" and have reported of the farmers' injury in hand in addition to the fatigue as 92.8% in upper-back, 93.6% in mid-back, and 91.8% in lower-back, respectively. Moreover, they have suggested "safe and good fit in hand" as the design-criteria for hand-tools and recommended the tool handle lengths as 12.4 cm in addition to the diameters as 3 cm, respectively.

Jain et al. (2018) have made a study on the risk-factors for MSDs among the farmers performing manual-harvesting in Rajasthan of India. In order to get the most influencing factors for MSDs in different body-parts, they have used the logistic-regression methodology, and found the fingers, wrists, shoulders, and lower-back as the body-parts in which more than 50% workers reported of MSDs. Also, except the shoulder and neck, the major associated factor with MSDs in all body parts was reported as age. Continuous adoption of both bending and twisting postures during work was significantly associated with prevalence of lower back problem and the workers have judged both of these as the most problematic job factors resulting in pain and injury (Roffey et al., 2010; Merlino et al., 2003). The high prevalence of MSDs for workers were due to various repetitive awkward movements like frequently working with the flexed back area, lifting and carrying heavy loads, etc., and poor working postures such as squatting, kneeling, etc., sustained for long working hours (Jain et al., 2018; Dianat & Salimi, 2014). Hence, there is a need to find out the possible risk factors among the farmers associated with harvesting and other farming practices. Different studies in India have reported MSDs and work-related upper limb disorders occurrence rates for farmers at about 63% to 98% (Das et al., 2013; Das & Gangopadhyay, 2015; Dhara et al., 2015). Nikkhah et al. (2017) have reported that the labour input energy of cultivation, crop protection, and harvesting of rice were 674.6, 356.8 and 205.9 MJ/ha, and the total labor input cost was 22,860,000 IRRs/ha, respectively. Also, the MSDs were commonly reported in lower back, back, and knees of paddy workers. It was further recommended to invest and accelerate the rice farming mechanization process in "Guilan Province of Iran". For the sustainable improvement in the profit and agricultural yields, it is highly desirable to apply the best ergonomic practices in the farming sectors to reduce the MSDs of farm workers. A major risk factor in developing of MSDs has been reported as high workload (Roman-Liu, 2014). Brandl et al. (2017b) have suggested that for reducing the risk, a suitable way could be of workload reduction. The main influencing factors for workload among its function such as exerted force, postures in working and sequence of time has been reported as exerted force as well as postures in working.

3.3.1 Evaluation of MSDs Among the Farmers

Although, any agriculture-based cultivation process involved a variety of activities like land-preparations, seeding, weed-pulling, and harvesting in addition to threshing processes, respectively. However, in this study only harvesting activities performed by 477 farmers based on the data collected through the use of "standard nordic-questionnaires" including the work-setting's characteristics, and MSDs in body-parts by the agricultural tools and machinery. Consideration of all the 477 farmers was on the basis of number of responses obtained, although the questionnaires were distributed among 500 paddy farmers in "Bangaladesh, Nepal, Srilanka, Pakistan, Indonesia and Thailand" performing same operations during farming. The designed questionnaires were with "Yes or No" type, and the body-parameters that were considered to be affected with MSDs because of the physical stresses included: "Pains in Shoulders and Neck; Lower and Upper Back-Pains; Pains in Wrists/Hands; Pains in Elbows and Thighs; and Pains in Knees and Ankles"; respectively.

Considering the farmers' work-setting's characteristics shown in Table 3.2, higher-incidence of MSDs were found in "lower and upper back-pains in sitting-postures; pains in shoulders and neck in squatting-postures; pains in wrists/hands in standing-postures; pains in shoulders and neck while working more than 3 hours-per-day in addition to more than 4 hours-per-week; and also pains in shoulders and

Table 3.2 Farmers' work-setting's characteristics

Parameters		Number of responses				
		Pains in Shoulders and Neck	Lower and Upper Back-Pains	Pains in Wrists/Hands	Pains in Elbows and Thighs	Pains in Knees and Ankles
Type of work-posture(s)	Sitting	226	228	223	128	220
	Squatting	224	129	222	129	129
	Standing	221	129	229	129	122
Working-hours per-day	Less than or equal to 3	128	127	121	120	125
	More than 3	129	224	127	119	118
Working-days per-week	Less than or equal 4	129	221	129	120	120
	More than 4	229	214	226	220	120
Working-time(s) per-day	Prior to 8 A.M.	29	27	29	23	21
	Between 8 A.M. to 12 noon	29	24	29	16	21
	Later than 12 P.M.	126	129	123	29	29
Job nature	Repetitive or forceful	129	121	129	227	224
	Carriage of heavy-loads	129	229	129	119	217

Total number of farmers (n) =477.

neck, when the job nature was repetitive or forceful and involved carriage of heavy-loads"; respectively. Similarly, by considering the MSDs in body-parts by the use of agricultural tools and machinery (Table 3.3), maximum farmers reported of pains in elbows and thighs as well as pains in wrists/hands while performing harvesting activities manually and partial-manually. While, maximum farmers reported of pains in shoulders and neck when carrying or transporting the appliances either completely-manually and partial-manually with the help of animals.

Thus, it was seen that most of the body-parts of farmers were subjected to the prevalence of MSDs. However, the prevention of MSDs depends on recognizing the risk-factors by self-assessment, study of workplaces as well as the working-postures while performing various activities. It also becomes important to consider the incidence-rate of disorders in workplaces in addition to the exposures to different hazardous working and/or environmental conditions. Through the identification of the working groups at some particular risk, the modifications of both the physiological and psychological environments can be carried out.

Table 3.3 MSDs in body-parts by using agricultural tools and machinery

Parameters		Number of responses				
		Pains in Shoulders and Neck	Lower and Upper Back-Pains	Pains in Wrists/Hands	Pains in Elbows and Thighs	Pains in Knees and Ankles
Harvesting activities	Manual-utilizations of hand-tools like sickles, knifes, etc.	326	220	223	328	120
	Partial-manual by the utilizations of hand -operated machines	121	127	129	127	122
	Completely-mechanized	28	29	21	20	25
Appliances' carriage and transportation	Completely manual-operated	329	324	327	219	218
	Partial-manual by the utilizations of animals	229	122	129	120	120
	Completely-mechanized	29	25	29	23	21

Total number of farmers (n) =477.

3.4 Agri-Equipment and IoT to Resolve the Physical Stress

So far the fatigues or work-related stresses are concerned, a number of modern tools and agri-based equipment have been manufactured in view of the requirements of farmers to reduce their drudgery-levels. Moreover, with the addition of "Internet of Things (IoT)" features, different improvements are further augmented. For example, collection of crops in field is normally carried out manually with the help of sickles which differs in the shape and size with the region. Nag et al. (1988) have emphasized on the design parameters of nine types of sickles and found that the geometry in cutting edge effectively contributes in being implemented. A further alteration was proposed in the configuration of improved sickle. Mishra et al. (2013) have examined the weeding effectiveness of devices and weeding action performed by farming ladies utilizing traditional tools and the newly introduced tools. A higher field capacity of 0.07 ha/h was reported by the improved sickles as compared to the normal sickles. Verma et al. (2016) have found that the utilization of their modified serrated sickle was good in ergonomic aspects, suitable to ladies, decreased drudgery level and in this way was useful in improving the efficiency of ladies. Chavan et al. (2015) have developed a manually operated reaper requiring 20 man-h/ha.

Further, the cost of harvesting with this reaper was reported as Rs. 1250.4 /ha, which was much lower in comparison to the traditional method having harvesting cost as Rs 2000/ha. Meisner et al. (1997) have found that in comparison to a daily labourer in the field doing cutting and placing cereals, a reaper is 14 times more efficient. In view of minimizing the operational cost, a BAU self-propelled reaper running by diesel fuel with 5 hp. engine (Model: Z175F-1), was developed through REFPI project during '1997–2003' (Hossain, 2003). Garg and Sharma (1991) have examined a tractor operated reaper windrower for harvesting of rice as well as wheat with harvesting capacity of 0.4 ha/h, having vertical conveyor belts. Further, by the use of this unit, the requirement of labour was reported to be reduced by two third in comparison with manually harvesting by the use of sickles.

Nalawade et al. (2009) have made a Jowar reaper windrower operated by tractor with cutting width of 2.2 m having 3 conveyor-belts to convey the cut stalks easily. Aung et al. (2014) have made the performance evaluation of power reaper having actual cutting width of 1.2 m with manual harvesting by the use of sickles for harvesting rice. It was found that the actual field capacity in addition to labour requirements for the power reaper was 0.24 ha/h and 4 man-h/ha, whereas for manual harvesting it was 0.05 ha/h and 28 man-h/ha, respectively. Further, cutting cost and grain loss of power reaper were reported to be 67% and 0.5% lesser in comparison to manual harvesting. Shalini et al. (2015) have designed a self-propelled reaper for cutting the soybean plants up to two rows. The machine included a 3.5 hp. diesel engine, pulley and belt drive, and collection mechanism with a cutting bar. The engine power was transmitted to the cutter through pulley and belt arrangement, and scissoring type of motion was utilized for crop cutting. Further, it was reported that the labour requirement was 20% as compared to manual harvesting. Chieh-Tsung (2012) have improved brush cutter operated by internal combustion engine into electric brush cutter having a DC motor as power source. Additionally, a Li-ion battery and an electronic control board for speed control of DC motor, as well as circuit protections were used. Vora et al. (2015) have reported of 63.30% and 43.84% financial savings in cost by using mini-tractor mounted reaper and self-propelled reaper, over the conventional method of harvesting crops by sickle. Mechanization of agricultural activities increases the labor productivity and crop intensity, as well as help to reduce the cost. It also helps in better farm management by providing more free time for next crop planting. Hossain et al. (2015) have revealed of savings about 97.50% of average time, 35.00% of cost and 2.75% of grains, by using combine harvester over manual methods. Noby et al. (2018) have reported of better technical and economic performances of a modified BAU self-propelled reaper than the existing BAU self-propelled reaper. During harvesting operation, the straw cutting performance of the reaper was not affected at different cutting speeds, whereas the throwing performances gradually increased with the increase in the ratio of cutting and forward speeds.

Different studies have been carried out in context with the application of IoT in farming sectors. For instance, Yan-e (2011) has discussed the agricultural information management concept in addition to the design method & architecture of "Intelligent Agriculture MIS". Vasisht et al. (2017) have discussed about

Farm-Beats's system design implementing an IoT platform for data collection by the use of different sensors, cameras along with drones in agriculture. Jawad et al. (2017) have made a review on the functions of Wireless Sensor Networks in agriculture in addition to the classification and comparison of different wireless communication protocols, and the energy harvesting techniques that can be used in agriculture-based monitoring systems. Lakshmi and Gayathri (2017) have combined image processing as well as IoT for monitoring the plant and for collecting the environmental factors such as temperature and humidity. A recognition system has been developed in image processing with the capability to identify plants by the use of the leaves images, such that the use of pesticides can be controlled with the help of the images. Different features which were extracted and compared included colors, textures and shapes of the leaves. Moreover, the pattern recognition helped to identify a leaf as deceased or healthy. Smart Sustainable Agriculture (SSA) includes "Smart Irrigation, Precision Farming, and Smart Greenhouse. Smart Irrigation refers to the improvement in efficiency of irrigation process and reduction in water losses though conservation of existing water resources by the use of smart irrigation systems based on IoT (Patrício & Rieder, 2018). The management of farms and conservation of resources by the use of IoT and 'Information & Communication Technologies (ICT)' are included in Precision Farming, which obtains real-time data pertaining to the condition of farm elements such as crops, soil and air, in order to protect the environment by ensuring profits as well as sustainability (Mohanraj et al., 2016). While, smart greenhouse promotes in cultivating crops with least human interventions by the use of continuous monitoring of climate conditions like humidity, luminosity, temperature and soil moisture, and then by automated triggering actions based on the changes evaluated, followed by corrective actions to maintain the favourable growing conditions (Bibri, 2018). Zhai et al. (2018) have proposed a "Precision-Farming System (PFS)" as a "Multi-Agent System (MAS)". The mission planning in PFS has been considered with several criterions such as energy-consumption, expected-benefits, or equipment-losses. An improved algorithm called as MP-PSOGA, has been proposed to solve the "Multi-objective Optimization Problem (MOP)" in mission planning, by taking help of the "Genetic Algorithm and Particle Swarm Optimization". Further, by performing a simulation called as precise pesticide spraying mission, the feasibility of the proposed approach was verified to ensure proper working of the proposed approach.

Dharmaraj and Vijayanand (2018) have proposed the application of "Artificial Intelligence (AI)" in the farming sectors such that farming quality can be enhanced, and also a quicker go-to-market strategy (GTM) strategy can be ensured for crops. The IoT nodes in agriculture includes small "Wireless-Sensor Networks (WSNs)" for monitoring that is similar as used in Building Automation, Process control, Security Systems, and Industries. Vaitheeka and Kumar (2018) have revealed that in smart farming with the implementation of IoT, the farmers can monitor the sensors to detect soil moisture, crop growths, pest-managements along with livestock tracking, and can improve the productivity with deduction in cost and also even in remote, can monitor the field. The IoT sensor nodes powered by solar energy were proposed for monitoring as well as controlling the agriculture fields (Sharma et al. (2019).

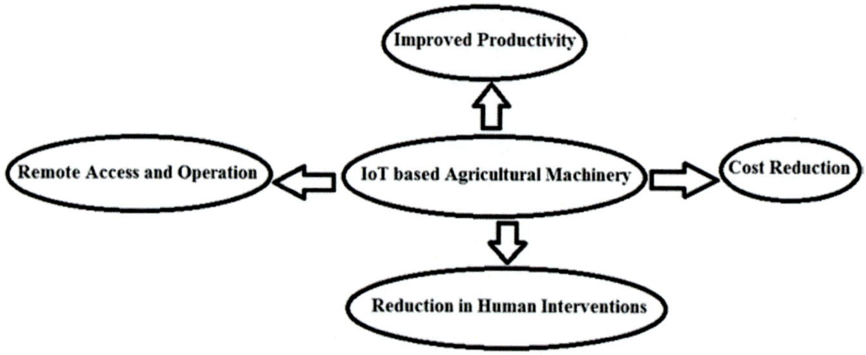

Fig. 3.2 Benefits of IoT-based agricultural machinery

Nóbrega et al. (2019) have discussed on IoT/Machine-2-Machine interaction that describes the design and deployment addressing the Sheep-IT service-requirements. Figure 3.2 illustrated the benefits of IoT-based agricultural machinery.

3.4.1 Developing IoT-Based Equipment to Reduce the Physical Stress of Farmers

With regard to the existing cultivation practices, the benefits of IoT utilization in agricultural sectors in the developing as well as under-developed countries require adequate attention. In view of the earlier studies made by Bull et al. (2016), Hatefi (2018), Shefat et al. (2018), and Tugrul and Cimen (2016), the "strength, weaknesses, opportunities as well as threats (SWOT)" analysis was used in this study for the evaluation of the potential-benefits of utilizing IoT in agriculture, and for these two cases were considered such as the "conventional agriculture-based machinery" and the "IoT-based agricultural machinery". In view of the deficiencies in the conventional agriculture-based machinery that may lead to the occurrences of physical as well as mental stress among the farmers in the developing countries, the fuzzy model was created for the physical stress as output and considering the deficiencies in the conventional agriculture-based machinery as inputs by using fuzzy logic in "MATLAB Version-2013".

The choice of fuzzy logic was on the basis of its successful applications in earlier studies with a variety of areas, which included in evaluating pain-intensity rating-scales (Araujo & Miyahira, 2009), soils (Gruijter et al., 2011; Rodrígueza et al., 2016), life-cycle impact (Feri & Hong-Chao, 2014), crops and agriculture-based lands (Murmu & Biswas, 2015), sustainability evaluation (Houshyar et al., 2017), wrist's diagnosis and treatment (Amiri & Khadivar, 2017), and so on. Further, based on the existing harvesting practices and procedures, which was followed by "discussing with the farmers as well as with agricultural equipment's manufacturers, and consulting with the experts", economical and compact IoT-based harvesting equipment was designed and developed in India. Then, its cutting efficiency in farming field was calculated and the opinion of farmers was obtained for both the existing and proposed equipment.

3.4.1.1 SWOT Analysis

With the use of SWOT analysis for evaluating the benefits of utilizing IoT in agriculture by considering two cases such as the "conventional agriculture-based machinery" and the "IoT-based agricultural machinery", it was found of the IoT-based machinery to be more suitable and useful for agricultural activities (Table 3.4).

Table 3.4 SWOT analysis of the "conventional agriculture-based machinery" and the "IoT-based agricultural machinery"

Conventional agriculture-based machinery	Strength	• Largely available • Accessible in wider range • Involvement of local manufacturers to design and fabricate smaller tools and equipment
	Weaknesses	• High initial as well as depreciation cost for large and heavy equipment • Skilled and trained operators requirement for large and heavy equipment
	Opportunities	• Friendly to most of the farmers • Higher productivity as compared to manual methods • Reduced time of operation
	Threats	• More fuel consumption for bigger machinery • Small-scale farmers find difficulties in affording to purchase the machinery
IoT-based agricultural machinery	Strength	• Requirement of only initial training so that lesser skilled individuals can operate easily • Higher efficiency and productivity as compared to existing agricultural machinery • Reduced weight as compared to existing agricultural machinery • Reduced human-intervention • Noiseless operations
	Weaknesses	• Reduced awareness among farmers and the farming communities in the developing or lesser developed countries • Lesser governmental supports and subsidies
	Opportunities	• Operable in remote-areas • Reduced efforts to operate • Higher coverage areas in less time • Enhanced incomes to the farmers
	Threats	• Higher initial costs for more sophisticated machinery • Higher training as well as awareness requirement among farmers and the farming communities

The strengths associated with the IoT-based machinery were included with "Requirement of only initial training so that lesser skilled individuals can operate easily; Higher efficiency and productivity as compared to existing agricultural machinery; Reduced weight as compared to existing agricultural machinery; Reduced human-intervention; and Noiseless operations". While the threats associated with the IoT-based machinery were included with "Higher initial costs for more sophisticated machinery; and Higher training as well as awareness requirement among farmers and the farming communities", respectively. However, the opportunities in comparison to the conventional agriculture-based machinery were revealed as "Operable in remote-areas; Reduced efforts to operate; Higher coverage areas in less time; and Enhanced incomes to the farmers".

3.4.1.2 Fuzzy Logic Modeling for Physical Stress Among Farmers

With the involvement of uncertainty, a fuzzy logic normally describes an elusiveness or vagueness in order to help in decision-making and/or controlling that requires no detailed mathematical-description for the scientific-tool for the system's simulation using fuzzy inference system. During fuzzification by using different required membership functions, the conversion of real values into fuzzy forms take place. Further, by including a set of "IF-THEN" linguistic-rules in the rule bases, the system's logical evolution based on the principal character's linguistic-values are obtained (Li, 2013). Thus, the use of fuzzy logic in this study was to create a fuzzy model for the evaluation of physical stress as output and considering the deficiencies in the conventional agriculture-based machinery as inputs. However, based on in-depth review of literature and subsequent experts' views, the dominant-factors related to the deficiencies in the conventional agriculture based machinery leading to both physical and mental stress among the farmers were obtained, which included four most dominant-factors such as: "Technical aspects (Hamsa & Bellundagi, 2017); Situational aspects (Romer Løvendal & Knowles, 2005); Welfare aspects (Romer Løvendal & Knowles, 2005); and Individual aspects (Hamsa & Bellundagi, 2017)"; respectively.

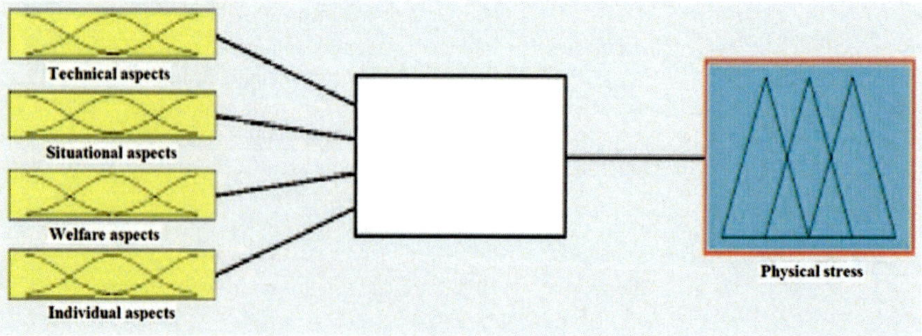

Fig. 3.3 Mamdani type fuzzy model with four inputs and one output

Further, the four obtained most dominant factors were taken as four inputs and physical stress as output creating a Mamdani type fuzzy model shown in Fig. 3.3. The input-data's fuzzification was obtained by using linguistic-variables. The range value's membership-functions were assigned with the values as follows: Technical aspects in the range of '0 to 3' representing 'Nil, Low, Medium, and High'; Situational aspects in the range of '0 to 3' representing 'Nil, Low, Medium, and High'; Welfare aspects in the range of '0 to 3' representing 'Nil, Low, Medium, and High'; and Individual aspects in the range of '0 to 3' representing 'Nil, Low, Medium, and High'; respectively. Similarly, the physical stress as output was assigned in the range between '0 and 1' representing '0=No and 1=Yes', respectively. Moreover, based on existing literature and experts' opinion, trapezoidal membership-functions were selected for all the four inputs as well as for the output.

Both "Fig. 3.4 and Fig. 3.5" illustrated the "rule base and fuzzy inference system" that were established in conformity to the defined rules for the model. Similarly, Fig. 3.6 illustrated the surface that was generated for evaluating the physical stress on the basis of the considered inputs.

3.4.1.3 Proposed Equipment

The proposed equipment was designed with a rotary plate cutter operated by DC motor at the front for performing the harvesting operations (Fig. 3.7). As rotary cutter was used, it can also perform the cutting operations in the left as well as right sides. With the help of load-cells connected with the handles, the pushing as well as pulling force was measurable in order to enable the operators about the continuity of their force-levels, such that they can take adequate measures by assessing themselves. Further, with the help of buzzers connected at the front, the operators were

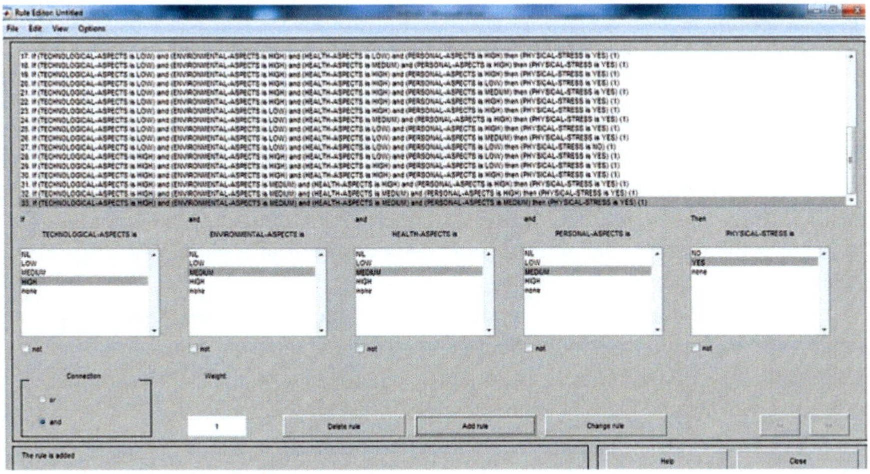

Fig. 3.4 The established model's rule base

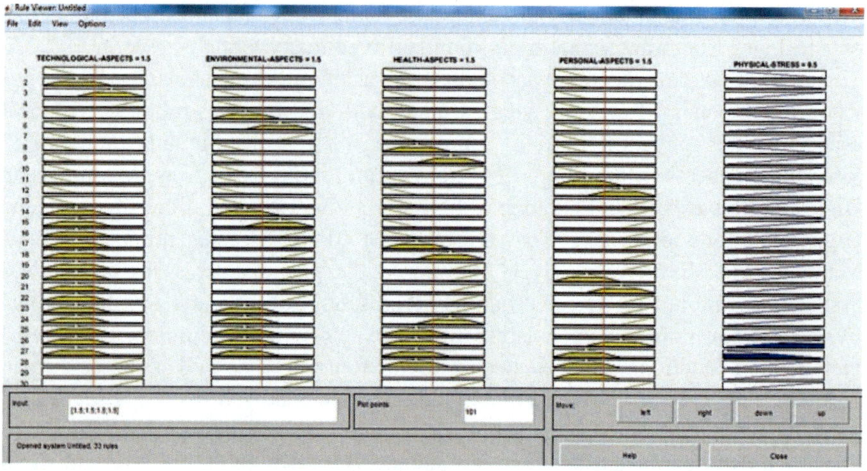

Fig. 3.5 The established model's MATLAB rules

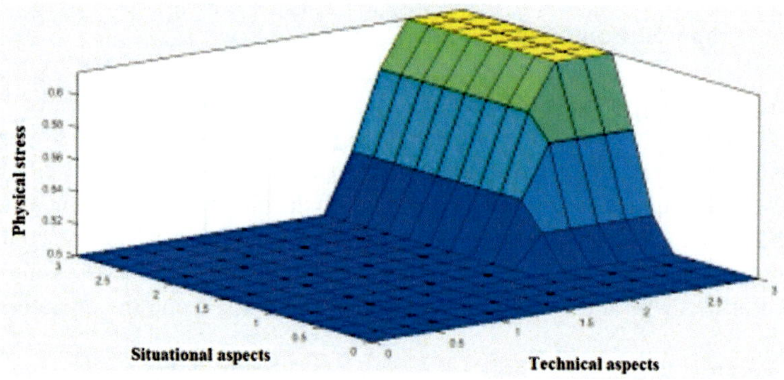

Fig. 3.6 The established model's generated surface

able to adjust the depth of cutting to the pre-determined levels. The whole unit was designed to run on battery powers, which saved the unnecessary costs associated with fuels. Further, the equipment was designed to run by operating a remote that reduced the drudgery and fatigue levels of operators in pulling or pushing. In order to achieve the required features, different sensors with programmable chips were attached accordingly.

As suggested by Hunt (1995), the performance indicators for the proposed equipment based on trial field treatments were obtained as follows: "Effective field-capacity =0.017ha/hr; Theoretical field-capacity =0.025ha/hr; Field-efficiency =68%; and Cutting-efficiency =99.91%"; respectively.

Specifications	
Parameters	**Proposed equipment**
Type of equipment	Remote-operated self-propelled as well as manual push-pull type
Overall dimensions (length×breadth×height) in m	1.45×0.48×0.85
Cutting height in m	0.10-0.20
Cutting width in m	0.35
Cutting tool	Rotating teethed cutter

Fig. 3.7 Proposed equipment

3.5 Conclusion

For the sustainable improvement in the profit and agricultural yields, it is highly desirable to apply the best ergonomic practices in the farming sectors to lessen the risk of musculoskeletal disorders (MSDs) of farmworkers. A major risk factor in developing of MSDs has been reported as high workload (Roman-Liu, 2014). Brandl et al. (2017a) have suggested that for reducing the risk, a suitable way could be of workload reduction. The agricultural tools and equipment, if designed properly by keeping the ergonomic aspects and they will be automated for most of its operations, then it would reduce the drudgery-level of farmers to considerable amount. This in-turn will be helpful in reducing the physical stresses as well as development of mental-stress among the farmers.

Moreover, the machines become smarter by implementing "Internet of Things (IoT)" technologies that enables intelligent data processing and make communication more effective and efficient. As IoT is an integration of a variety of things or devices, for instance radio frequency identification, sensors, mobile phones, actuators, drones, etc. for communicating with each other for the accomplishment of some common goal. Therefore, it is useful in changing a wide-range of instantaneous monitoring applications in agricultural automation. Also, IoT is a novelty approach in the wireless communication field, which involves many intelligent agents sharing information to enable collaborative decisions making in addition to the accomplishment of tasks in an optimal way.

References

Amiri, F. M., & Khadivar, A. (2017). A fuzzy expert system for diagnosis and treatment of musculoskeletal disorders in wrist. *Technical Gazette, 24*(Suppl.1), 147–155.

Araujo, E., Miyahira, S. A. (2009) Unidimensional fuzzy pain intensity scale. IEEE International Conferences on Fuzzy Systems pp.185–190.

Aung, N. N., Myo, P. P., & Moe, H. Z. (2014). Field performance evaluation of a power reaper for paddy harvesting. *International Journal of Scientific Engineering and Technology Research, 12*(3), 2631–2636.

Bhattacharyya, N., & Chakrabarti, D. (2012). Ergonomic basket design to reduce cumulative trauma disorders in tea leaf plucking operation. *Work, 41*, 1234–1238.

Bibri, S. E. (2018). The IoT for smart sustainable cities of the future: An analytical framework for sensor-based big data applications for environmental sustainability. *Sustainable Cities and Society, 38*, 230–253.

Brandl, C., Mertens, A., & Schlick, C. M. (2017a). Effect of sampling interval on the reliability of ergonomic analysis using the Ovako working posture analysing system (OWAS). *International Journal of Industrial Ergonomics, 57*, 68–73. https://doi.org/10.1016/j.ergon.2016.11.013

Brandl, C., Mertens, A., & Schlick, C. M. (2017b). Ergonomic analysis of working postures using OWAS in semi-trailer assembly, applying an individual sampling strategy. *International Journal of Occupational Safety and Ergonomics, JOSE*, 1–8. https://doi.org/10.1080/10803548.2016.1191224

Bull, J. W., Jobstvogt, N., Böhnke-Henrichs, A., Mascarenhas, A., Sitas, N., Baulcomb, C., Lambini, C. K., Rawlins, M., Baral, H., Zähringer, J., Carter-Silk, E., Balzan, M. V., Kenter, J. O., Häyhä, T., Petz, K., & Koss, R. (2016). Strengths, weaknesses, opportunities and threats: A SWOT analysis of the ecosystem services framework. *Ecosystem Services, 17*, 99–111. https://doi.org/10.1016/j.ecoser.2015.11.012

Cakmak, B., Saracoglu, T., Alayunt, F. N., & Ozarslan, C. (2011). Vibration and noise characteristics of flap type olive harvesters. *Applied Ergonomics, 42*, 397–402.

CDC. (2013). *Agricultural safety*. Centers for Disease Control and Prevention. Retrieved from www.cdc.gov/niosh/topics/aginjury/

Chaturvedi, V., Kumar, A., & Singh, J. K. (2012). Power tiller: Vibration magnitudes and intervention development for vibration reduction. *Applied Ergonomics, 43*, 891–901. https://doi.org/10.1016/j.apergo.2011.12.012

Chavan, P. B., Patil, D. K., & Dhondge, D. S. (2015). Design and development of manually operated reaper. *IOSR Journal of Mechanical and Civil Engineering, 12*(3), 15–22.

Chieh-Tsung, C. (2012). A new electric brush cutter. *WSEAS Transactions on Systems and Control, 3*(7), 2224–2856.

Copuroglu, C., Heybeli, N., Ozcan, M., Yilmaz, B., Ciftdemir, M., & Copuroglu, E. (2012). Major extremity injuries associated with farmyards accidents. *The Scientific World Journal, 314038*, 6. https://doi.org/10.1100/2012/314038

Das, B. (2014). Assessment of occupational health problems and physiological stress among the brick field workers of West Bengal, India. *International Journal of Occupational Medicine and Environmental Health, 27*(3), 413–425.

Das, B. (2015). An evaluation of low back pain among female brick field workers of West Bengal, India. *Environmental Health and Preventive Medicine, 20*(5), 360–368.

Das, B., & Gangopadhyay, S. (2011). An ergonomics evaluation of posture related discomfort and occupational health problems among paddy farmers. *Occupational Ergonomics, 10*(1, 2), 25–38.

Das, B., & Gangopadhyay, S. (2015). Prevalence of musculoskeletal disorders and physiological stress among adult, male potato cultivators of West Bengal. *India. Asia Pacific Journal of Public Health, 27(2):NP1669-NP1682.*

Das, B., Ghosh, T., & Gangopadhyay, S. (2013). Child work in agriculture in West Bengal, India: Assessment of musculoskeletal disorders and occupational health problems. *Journal of Occupational Health*, 12–0185.

Dhara, P. C., De, S., Sengupta, P., Maity, P., & Pal, A. (2015). An ergonomic approach for designing indian traditional vegetable cutter. *Work, 50*(2), 177–186.

Dharmaraj, V., & Vijayanand, C. (2018). Artificial Intelligence (AI) in Agriculture. *International Journal of Current Microbiology and Applied Sciences, 7*(12), 2122–2128.

Dianat, I., & Salimi, A. (2014). Working conditions of Iranian hand-sewn shoe workers and associations with musculoskeletal symptoms. *Ergonomics, 57*(4), 602–611.

Dooley WK (2012) Ergonomics and the development of agricultural vehicles, agricultural equipment technology conference.

Dube, B., Mishra, D., Satapathy, S., & Singh, S. (2018). Ergonomic evaluation of seeding, fertilizing and weeding postures in agricultural field. *International Journal of Pure and Applied Mathematics, 119*(10 Special Issue B), 1875–1880.

Fathallah, F. A. (2010). Musculoskeletal disorders in labor-intensive agriculture. *Applied Ergonomics, 41*(6), 738–743.

Feri, A., & Hong-Chao, Z. (2014). A fuzzy logic based aggregation method for LCIA. *Journal of Cleaner Production, 67*, 159–172.

Garg, I. K., & Sharma, V. K. (1991). Tractor-operated vertical-conveyer reaper windrower. *Indian Farming, 40*(11), 32–36.

Giuffrida, A., Iunes, R. F., & Savedoff, W. D. (2001). *Economic and health effects of occupational hazards in Latin America and the Caribbean*. Inter-American Development Bank. Retrieved from http://services.iadb.org/wmsfiles/products/Publications/353816.pdf

Gruijter, J., Walvoort, D. J., & Bragato, G. (2011). Application of fuzzy logic to boolean models for digital soil assessment. *Geoderma, 166*, 15–33.

Hamberg-van Reenen, H. H., van der Beek, B. M., Blatter, A. J., van der Grinten, M. P., van Mechelen, W., & Bongers, P. M. (2008). Does musculoskeletal discomfort at work predict future musculoskeletal pain? *Ergonomics, 51*(5), 637–648.

Hamsa, K. R., & Bellundagi, V. (2017). Review on decision-making under risk and uncertainty in agriculture. *Economic Affairs, 62*(3), 447–453. https://doi.org/10.5958/0976-4666.2017.00056.0

Hatefi, S. M. (2018). Strategic planning of urban transportation system based on sustainable development dimensions using an integrated SWOT and fuzzy COPRAS approach. *Global Journal of Environmental Science and Management, 4*(1), 99–112. https://doi.org/10.22034/gjesm.2018.04.01.010

Hellig, T., Mertens, A., & Brandl, C. (2018). The interaction effect of working postures on muscle activity and subjective discomfort during static working postures and its correlation with OWAS. *International Journal of Industrial Ergonomics, 68*, 25–33. https://doi.org/10.1016/j.ergon.2018.06.006

Hossain MD (2003) *Fabrication of self-propelled reaper by locally available materials*. MS thesis, Department of Farm Power and Machinery, BAU, Mymensingh, Bangladesh.

Hossain, M. A., Hoque, M. A., Wohab, M. A., Miah, M. A. M., & Hassan, M. S. (2015). Technical and economic performance of combined harvester in farmers' field. *Bangladesh Journal of Agricultural Research, 40*(2), 291–304.

Houshyar, E., Smith, P., Mahmoodi-Eshkaftaki, M., & Azadi, H. (2017). Sustainability of wheat production in Southwest Iran: A fuzzy-GIS based evaluation by ANFIS. *Cogent Food & Agriculture, 3*, 1327682. https://doi.org/10.1080/23311932.2017.1327682

Hunt, D. (1995). *Farm power and machinery management, cost determination* (9th ed.). Iowa State University Press.

Jain, R., Meena, M. L., & Dangayach, G. S. (2016). Need of agriculture hand tool design using quality and ergonomics principles. In *CAD/CAM, robotics and factories of the future* (pp. 77–84). Springer.

Jain, R., Meena, M. L., Dangayach, G. S., & Bhardwaj, A. K. (2018). Risk factors for musculoskeletal disorders in manual harvesting farmers of Rajasthan. *Industrial Health, 56*(3), 241–248.

Jatinder, K. J., & Rana, A. (2012). Ergonomic interventions in weeding operations for drudgery reduction of hill farm women of India. *Work, 41*, 4349–4355.

Jawad, H. M., Nordin, R., Gharghan, S. K., Jawad, A. M., & Ismail, M. (2017). Energy-efficient wireless sensor networks for precision agriculture: A review. *Sensors, 17*, 1781. https://doi.org/10.3390/s17081781

Karhu, O., Kansi, P., & Kuorinka, I. (1977). Correcting working postures in industry: A practical method for analysis. *Applied Ergonomics, 8*(4), 199–201. https://doi.org/10.1016/0003-6870(77)90164-8. PMID:15677243.

Karhu, O., Harkonen, R., Sorvali, P., & Vepsalainen, P. (1981). Observing working postures in industry: Examples of OWAS application. *Applied Ergonomics, 12*(1), 13–17. https://doi.org/10.1016/0003-6870(81)90088-0. PMID:15676393.

Kumar, A., Singh, J. K., Mohan, D., & Varghese, M. (2008). Farm hand tools injuries: A case study from northern India. *Safety Science, 46*, 54–65. https://doi.org/10.1016/j.ssci.2007.03.003

Lakshmi, K., & Gayathri, S. (2017). Implementation of IoT with image processing in plant growth monitoring system. *Journal of Scientific and Innovative Research, 6*(2), 80–83.

Li, Q. (2013). A novel Likert scale based on fuzzy sets theory. *Expert Systems with Applications, 40*, 1609–1618.

MATLAB. (2013). The MathWorks, Inc., Natick, Massachusetts, United States.

McAtamney, L., & Corlett, E. N. (1993). RULA: A survey method for the investigation of work related upper limb disorders. *Applied Ergonomics, 24*(2), 91–99.

Meisner, C. A., Petter, H., Badruddin, M., Razzaque, M. A., Giri, G. S., Scott, J. (1997). *Mechanical revolution among small landholders of South Asia: The growing use of Chinese hand tractors*. Proceedings of the Joint International Conference on Agricultural Technology and Engineering Exhibition '97, Dhaka 3:781–787.

Merlino, L. A., Rosecrance, J. C., Anton, D., & Cook, T. M. (2003). Symptoms of musculoskeletal disorders among apprentice construction workers. *Applied Occupational and Environmental Hygiene, 18*(1), 57–64.

Mishra, D., & Satapathy, S. (2018a). Drudgery reduction of farm women of Odisha by improved serrated sickle. *International Journal of Mechanical Engineering and Technology, 9*(2), 53–61.

Mishra, D., & Satapathy, S. (2018b). Hand tool injuries of agricultural farmers of South Odisha in India. *Materials Today: Proceedings, 5*(9), 17648–17653.

Mishra, R., Singh, Y. P., Mishra, Y. D., Singh, S., & Singh, H. (2013). Dissemination of improved sickles for female agriculture workers for crop harvesting. *A Journal of Multidisciplinary Advance Research, 2*(1), 118–123.

Mohanraj, I., Ashokumar, K., & Naren, J. (2016). Field monitoring and automation using IOT in agriculture domain. *Procedia Computer Science, 93*, 931–939.

Murmu, S., & Biswas, S. (2015). Application of fuzzy logic and neural network in crop classification: A review. *Aquatic Procedia, 4*, 1203–1210.

Naeini, H. S., Karuppiah, K., Tamrin, S. B., & Dalal, K. (2014). Ergonomics in agriculture: An approach in prevention of work-related musculoskeletal disorders (WMSDs). *Journal of Agriculture and Environmental Sciences, 3*, 33–51.

Nag, P. K., Goswami, A., Ashtekar, S. P., & Pradhan, C. K. (1988). Ergonomics in sickle operation. *Applied Ergonomics, 19*(3), 233–239.

Nalawade, S. M., Turbatmath, P. A., & Gajakos, A. V. (2009). Design and development of tractor operated Jowar reaper windrower. *New Agriculturist, 20*(1,2), 75–81.

Nikkhah, A., Kougir-Chegini, Z., Kosari-Moghadam, A., & Nikkhah, F. (2017). Musculoskeletal disorders, energy use, and costs of human labor for Paddy cultivation in northern Iran. *International Journal of Agricultural Management and Development, 7*(4), 439–446.

Nóbrega, L., Gonçalves, P., Pedreiras, P., & Pereira, J. (2019). An IoT-based solution for intelligent farming. *Sensors, 19*, 603. https://doi.org/10.3390/s19030603

Noby, M. M., Hasan, M. K., Ali, M. R., Saha, C. K., Alam, M. M., & Hossain, M. M. (2018). Performance evaluation of modified BAU self-propelled reaper for paddy. *Journal of Bangladesh Agricultural University, 16*(2), 171–177. https://doi.org/10.3329/jbau.v16i2.37956

Nunes, I. L. (2009). FAST ERGO_X – A tool for ergonomic auditing and work-related musculoskeletal disorders prevention. *WORK: A Journal of Prevention, Assessment, & Rehabilitation, 34*(2), 133–148.

Osborne, A., Blake, C., Fullen, B. M., Meredith, D., Phelan, J., McNamara, J., & Cunningham, C. (2012). Prevalence of musculoskeletal disorders among farmers: A systematic review. *American Journal of Industrial Medicine, 55*(2), 143–158.

Parvez, M. S., & Shahriar, M. M. (2018). Agricultural farm-related injuries in Bangladesh and convenient design of working hand tools. *Journal of Healthcare Engineering*, 2018.

Patel, J., & David, S. (2016). Ethics in occupational health and safety: Case studies from Gujarat. *Indian Journal of Medical Ethics, 1*(4), 203–210.

Patel, S. K., Varma, M. R., & Kumar, A. (2010). Agricultural injuries in Etawah district of Uttar Pradesh in India. *Safety Science, 48*(2), 222–229.

Patrício, D. I., & Rieder, R. (2018). Computer vision and artificial intelligence in precision agriculture for grain crops: A systematic review. *Computers and Electronics in Agriculture, 153*, 69–81.

Pyykkönen, M., & Aherin, B. (2012). Chapter 53: Occupational health and safety in agriculture. In *Sustainable agriculture* (pp. 391–401). Uppsala University.

Ren, J., & Xiao, D. (2009). Ergonomic simulation and evaluation of elliptical trainer with CATIA V5. *CADDM, 19*(1), 81–86.

Rodrígueza, E., Pechea, R., Garbisub, C., Gorostizac, I., Epeldeb, L., Artetxed, U., Irizare, A., Sotoe, M., & Mabecerrild, J. (2016). Dynamic quality index for agricultural soils based on fuzzy logic. *Ecological Indicators, 60*, 678–692.

Roffey, D. M., Wai, E. K., Bishop, P., Kwon, B. K., & Dagenais, S. (2010). Causal assessment of occupational sitting and low back pain: Results of a systematic review. *The Spine Journal, 10*(3), 252–261.

Roman-Liu, D. (2014). Comparison of concepts in easy-to-use methods for MSD risk assessment. *Applied Ergonomics, 45*(3), 420–427. https://doi.org/10.1016/j.apergo.2013.05.010. PMID:23849897.

Romer Løvendal, C., & Knowles, M. (2005). *Tomorrow's hunger: A framework for analysing vulnerability to food insecurity* (ESA Working Paper No. 05-07). FAO.

Sanjog, J., Karmakar, H., Agarwal, C., & Patil, D. (2012). Designing and ergonomic evaluation of a shoe-rack in CAD environment. *International Journal of Computer Applications, 49*(20), 38–41.

Shalini, P. S., Wankhade, A. M., & Syed, M. N. (2015). Design and fabrication of crop reaper. *International Journal for Engineering Applications and Technology*. ISSN:2321-8134.

Sharma, H., Haque, A., & Jaffery, Z. A. (2019). Smart agriculture monitoring using energy harvesting internet of things (EH-IoT). *World Scientific News, 121*, 22–26.

Shefat, S. H. T., Rahman, A., Chowdhury, M. D. A., & Uddin, M. N. (2018). Strength, weakness, opportunities and threat analysis of integrated aqua-farming in Bangladesh. *Acta Scientific Agriculture, 2*(12), 112–118.

Soheili-Fard, F., Rahbar, A., & Marzban, A. (2017). Ergonomic investigation of workers in tea factories using REBA and OWAS methods -case study: (Langroud region, Guilan, Iran). *Agricultural Engineering International: CIGR Journal, 19*(3), 112–119.

Tabibi, R., Ghorbani-Kalkhajeh, S., Tarahomi, S., Valipour, A. A., Tajzadeh, S., & Sokooti, M. (2017). Occupational health surveillance programs for Iranian farmers. *Arvand Journal of Health Medicine Science, 2*(2), 37–39. https://doi.org/10.22631/ajhms.2017.86398.1018

Tugrul, B., Cimen, S. (2016). *Importance of corporate governance for energy in sustainable development and evaluation with quantitative SWOT analysis*. Special issue of the 2nd international conference on computational and experimental science and engineering (ICCESEN 2015), Acta Physica Polonica A:130(1).

Vaitheeka, N., & Kumar, S. M. (2018). Cognitive intelligence of internet of things (IoT) in agriculture industry-a primer study on smart farming. *International Journal of Pure and Applied Mathematics, 118*(20), 1835–1841.

Vasisht, D., Kapetanovic, Z., Won, J. H., Jin, X., Chandra, R., Kapoor, A., Sinha, S. N., Sudarshan, M., Stratman, S. (2017). *FarmBeats: An IoT platform for data-driven agriculture*. 14th USENIX Symposium on Networked Systems Design and Implementation (NSDI '17), Boston, MA, USA.

Verma, S., Gupta, S., & Pachauri, C. P. (2016). Superiority of the Naveen serrated sickle over the traditional sickle for wheat harvesting. *International Journal of Farm Sciences, 6*(1), 214–222.

Villarejo, D., & Baron, S. L. (1999). Occupational health status of hired farm workers. *Occupational Medicine, 14*, 613–635.

Vora, M. D., Parmar, R. P., & Gunjaria, H. M. (2015). Comparative performance evaluation of Paddy harvesting. *Trends in Biosciences, 8*(19), 5411–5416.

Vyavahare, R. T., & Kallurkar, S. P. (2012). Anthropometric and strength data of Indian agricultural workers for equipment design: A review. *Agricultural Engineering International: CIGR Journal, 14*(4), 102–114.

White, G., & Cessna, A. (1989). Occupational hazards of farming. *Canadian Family Physician, 35*, 2331–2336.

Wibowo, R. K. K., & Soni, P. (2016). Farmers' injuries, discomfort and its use in design of agricultural hand tools: A case study from East Java, Indonesia. *Agriculture and Agricultural Science Procedia, 9*, 323–327.

Yan-e, D. (2011). *Design of intelligent agriculture management information system based on IoT*. Fourth International Conference on Intelligent Computation Technology and Automation, IEEE. https://doi.org/10.1109/ICICTA.2011.262.

Zhai, Z., Ortega, J.-F. M., Martínez, N. L., & Rodríguez-Molina, J. (2018). A Mission planning approach for precision farming systems based on multi-objective optimization. *Sensors, 18*, 1795. https://doi.org/10.3390/s18061795

Chapter 4
Measurement and Prediction of Mental Stress Through Innovative Equipment in Agriculture

4.1 Introduction

By exposing to adverse climatically-conditions while performing the farm-related activities, the whole farming-community put them in some kind of occupational-hazards affecting their health-condition by considerable reduction in the farm-productivity, work-quality and thus by increasing the musculoskeletal-related disorders. The workplace associated tasks are very harmful because of the physical and emotional responses of human-beings owing to varying situational aspects. Higher job-demands in addition to reduced controlling abilities of situations may cause stresses in workplaces. Regarding the farm-related occupation, there occurs more dangerous consequences with more rates of fatality as compared to some other occupation. Normally the farmers do their work for longer duration of time with different equipment that are untrustworthy and also, in challenging weather-condition with hazardous-chemical substances and so on. In addition, they also make large financial commitments exposing them to a higher financial risk-level. The human-beings now a day are more exposed to mental stress that develops and increases a number of problems including depressions, reduction of work ability, performance-levels, heart-attacks, suicidal-rates, and so on. Most of the farmers in particular work at different adverse climatic states under unfavorable situations, which result in higher depression-levels resulting in mental ill-health condition.

Owing to different existing situations in workplaces faced continuously by individuals such as work-overloads, job insecurity, job dissatisfaction, and work-pressures, the continuous presence of stresses occur resulting in mental stresses. The mental stress's presence not only affects the health as well as well-being of farm workers, but also affects their performance and productivity affecting the overall profit-margin. By the occurrence of these mental-related stresses, which cannot be handled effectively by the farmers, they are subjected to both mental as well as physical changes. Therefore, proper monitoring of the induced mental stress-levels

S. Satapathy et al., *Innovation in Agriculture with IoT and AI*,
Innovation, Technology, and Knowledge Management,
https://doi.org/10.1007/978-3-030-88828-2_4

has been a prime requirement to assess the current-state of mind of the human-beings for possible intervention-strategies to undertake. Thus, an attempt has been made in this study to explore the mental workload associated in the agricultural sectors, and possible innovative ways to measure and predict the occurrence of mental stress among the farmers as well as the farming-communities.

Based on the objectives of this study, the following research questions were formulated, such as: a) What are the causes and the factors that affect the mental workload in the form of mental stress among the farmers? and b) How can the mental stress be measured and predicted in order to reduce its adverse effects?

4.2 Mental Workload in Agri-Farming

As agriculture has been regarded as a hazardous-based occupation affecting the farm-worker's health, so the identification of the associated stressors can be helpful for the health-related professionals to improve the health-outcomes with significant intervention strategies (Kearney et al., 2014). The present modern society's most important issue is the day-to-day life stresses becoming an unavoidable part as well as growing issue. The psychological stresses may be either "acute or chronic" (Picard, 2016). However, the stresses related to work-activities in addition to the psycho-social risks have been revealed in the consequence of losses in working-days in case of business-sectors (Alberdi et al., 2016). The use of machine-learning to mental-health has resulted in a number of benefits including the areas such as: "diagnosis, treatments as well as support, different-researches and clinical-administrations" (Shatte et al., 2019). Mallikarjun et al. (2019) have stated that an unbalanced mental-state results in depressions damaging distresses of a someone's thinking and/or feeling and/or reaction. Throughout the world, the presence of stress has been regarded as most-significant health-issue. The adverse effect of different stresses on human-health include "emotional-distresses, muscle-oriented aches as well as tensions, pains in the back, headaches, heartburn, digestive-related tract-issues, and over-arousal" owing to the acute-stress occurrence (American Psychology Association, 2019).

For the evaluation of the work-related loads, the measurement of heart rate can be used in different cases. As the increased demands for oxygen by the muscles are always associated with increasing heart rates, its working condition influences the emotional in addition to the mental stress. Imbeau et al. (1995) determined the total stress by using the R-wave of "electro-cardiogram (ECG)". Previous studies have indicated that other than the musculoskeletal-related illness causing in stresses (Colligan & Higgins, 2006), an increased rate of stress-related diseases lead to absenteeism as well as staffs turn-over (Milczarek & Elke Schneider, 2009). Occurrences of more acute stress due to the prevalent stresses are more in comparison to the chronic stress (England et al., 2012). However, the occurrence of chronic stress is because of the pressure-levels of longer standing-postures and the socioeconomic based demands, and problems associated with the interpersonal-relationship leading to more serious health-related issues (England et al., 2012). Because of higher repetitive-activities, more load weights, and unfavorable work-postures, the farmers' musculoskeletal-based systems are normally subjected to higher-stresses that may have harmful-effect on their health-conditions leading to working incapacity of temporary or permanent natures (Rammelmeier et al., 2014).

4.3 Prediction of Mental Workload and Its Impact Due to Physical Stress

The use of fuzzy-based models has been increasing both in the industrial as well as academic areas and researches, mainly for data-analysis and system's control (Chakraborty & Barma, 2014; Gajjar et al., 2014; Pandey & Parhi, 2017; Ramanathan, 2014; Sawhney et al., 2014). However, fewer studies have been done regarding the application of fuzzy-environment for the agricultural farmers' mental-conditions related issues. Berkowitz and Perkins (1984) have considered 126 dairy-based farmwives for examining the stresses among them by utilizing "path-analysis" in order to assess the relative contributions to work-related stresses as well as for variables related to family. In a study, the translation of simple fuzzy-logic based model was done into a "Bayesian-based setting" by Lazaridis (1999) and further, the author addressed a 'discrimination and attribution/classification' problem.

Owing to the aggravation of a greater number of occupational-related stressors in the recent years, significant numbers of suicidal cases have been reported among farmers and farm-related workers in United Kingdom than any other occupational groups (Gregoire, 2002). Apart from the utilization of both "Sinha's Comprehensive Anxiety-Test" and "Approval Motive-Scale", Kushwaha and Kumar (2009) have also used traditional and fuzzy-based approaches for their psychology-based research. They proposed a stress-detection system based on physiological-signals, which used "heart-rates (HR) as well as galvanic skin-responses (GSR)" for providing information on the state of mind of individuals in view of the non-invasiveness as well as non-intrusiveness (de Santos, Ávila, Casanova, & del Pozo, 2011a; de Santos, Ávila, del Pozo, & Casanova, 2011b). A study has reported of the speech of persons to be reflecting the stress-levels or depression-levels (Gulhane et al., 2011). Both economic as well as natural constraints haw been reported as major-concerns for the farmers getting exposed to variety of stress-levels and constraints (Ghatul, 2013). A design framework was presented by Torres-Salomao et al. (2014) for the undergoing human-participants with mental stresses. For this, they utilized "Interval Type-2 Fuzzy-Logic (IT2FL)" architecture that was optimized with "Genetic-Algorithms" in addition to an efficient type "reduction-algorithm" for the modelling-technique. According to Salazar-Ramirez et al., the possibilities in the detection of any person's entry to the state of mental-blockages due to the occurrence of stresses can help in facilitation of possible mitigation strategies for the obstruction-symptoms. Quantification of mental stress by using fuzzy-analysis of "ECG-parameters" was done by Mittal and Grewal (2014). For this, they decomposed the 'ECG-signals' with "BIOR-3.9 wavelet-family" up to three-levels.

With the use of fuzzy-logic, El-Samahy et al. (2015) tried for automatic-modification of the induced mental stress based on a linguistic-model's output by employing fuzzy-sets in "Mamdani-type" fuzzy-rules' form. By considering 121 dairy-based farmers, Furey et al. (2016) have assessed their financial-related threats, social-related supports, anxieties, depressions, farm-related job stresses, health and safety-beliefs playing vital-roles in causing farm-related injuries. An investigation of 182 female farm-based workers was done by Cezar-Vaz et al. (2015) and found them with mental-related disorders along with moderate-levels of physical discomforts to be significant at high-level of working-time, physical-demands, frustrations and total efforts with regard to farm-related works, as well as significant at high-levels of frustrations in house-related works. By considering a total of 265 "Finnish dairy-based farmers", Kallioniemi et al. (2016) have identified the associated stressors in addition to the occurrence of stress, burnouts, and responsible variables for those symptoms. The stressors were found to be related to working load, and the health as associated with stress and burnouts' symptoms. Further, loneliness and poor economical situations were revealed as associated with stress. The occurrence of stresses is harmful to human-beings and thus, it becomes an essential requirement for the individuals to keep them fit (Padhy & Raju, 2018). In an investigation, Brigance et al. (2018) have found the mental-based health to be influenced in the working-place with 'organic practices' having some kind of risk factors as stated by non-organic and conventional farm workers. Hagen et al. (2019) have stated of very limited studies on the mental-health of farm-based communities, although these issues have been a major health-concern among the farm-based workers (Hovey & Seligman, 2006). Taouche (2019) have utilized the fuzzy-logic approach to predict the 'mechanical-parameters' including "stress-strain behavior-parameters" for the "α-phase as well as β-phase resistance" as a function of "α-phase volume-fraction". Daghagh Yazd et al. (2019) have considered 167 articles that were related to farmers' mental-health and also the key risk factors around the world affecting the farmers' mental-health. They found of four most-cited influencing risk factors, including "pesticide-exposures, finance-based difficulties, climatic-variability or droughts, and miserable physical-health or earlier-injuries. Usually, stress is the result of the reaction to changes in body in terms of actions in physical, mental, and emotional responses. Moreover, our body normally exposes to different stresses because of various events happening in addition to the performed things (Lakshmi et al., 2020).

For the increasing forecasting-related applications, the use of "artificial-intelligence based algorithms" has been found suitable not only for non-linear forecasting, but also in dealing with different complex-issues with involvement of uncertainties. Al-Hmouz et al. (2011) have used the "adaptive-network-based fuzzy inference-system (ANFIS)" in order to deliver adaptive-learning contents to mobile learners. By the use of ANFIS-models, a good match between predicted and the actual responses can be achieved with better-performance as compared to any linear and/or exponential regression-model. The use of ANFIS has been increased in the past for a variety of studies such as: for physical-properties of "tactile textures" (Akay et al., 2012); for analyzing the working-condition's effects on occupational-related injury (Fragiadakis et al., 2014); for the knowledge of the "architecture and basic-learning process of ANFIS" (Walia et al., 2015); for estimating the ice-creams' overall sensory-acceptance including other related-products (Bahram-Parvar et al., 2017); for attendance-rates' forecasting at 'soccer-games' (Şahin & Erol, 2017); for 'water quality-index' estimating (Tiwari et al., 2018);for product-distributions in multi echelon-transshipment systems (Okwu & Adetunji, 2018); for decision-support systems in medicals (Yadollahpour et al., 2018); for proposing of "maximum power-point tracking (MPPT) controller" with more accuracy (Koochaksaraei & Izadfar, 2019); and for the prediction of muscle-based strength of farm workers (Mishra & Satapathy, 2020).

4.3.1 Data Collection and Analysis

In order to obtain clarity on the relationship between physical stress and mental workload among the farmers, a designed questionnaire "based on experts' opinions and existing literature (Lemyre et al., 1990)" was distributed among them as illustrated in Table 4.1. For data collection, a total of 200 farmers were chosen as follows: 140 from India; 23 from Srilanka; and 37 from the Central Mexico; respectively. The selection of participants was based on their consent after clearly explaining the purpose of the work. Further, based on the reading obtained by the use of commercially available standard devices, the associated parameters such as "Blood-Pressure Levels (BPL) and Body-Temperature (BT)" with regard to the mental workload leading to mental stress, a fuzzy-model was developed in the fuzzy-environment for the mental stress as output and both BPL and BT as inputs with assigned range values by the use of "MATLAB-2013″ version software.

Moreover, in view of the wider applications as well as forecasting accuracy of ANFIS-based models, this work focused on identifying the mental stress of farmers owing to their measured associated parameters (BPL and BT) while performing the agricultural activities. Then, subsequent prediction of mental stress was done with regard to the parameters by creating ANFIS-based model and the choice of range

Table 4.1 Questionnaire

Variables	Responses Yes ()	No ()
I am feeling to be rushed without having enough-time to perform my tasks.		
I am getting difficulties to control my moods, emotions, reactions and/or gestures.		
I feel as if confused with my muddled thoughts and lacking in concentration to focus on work.		
I am getting physical-pains: Sore-back, headaches, stiff-neck, etc.		
I am feeling as if my shoulders are with great weights.		
I feel as if pre-occupied, worried or tormented because work-pressures.		
I am feeling of stressed due to work-pressures.		

values (0 = No and 1 = Yes) for the mental stress as output were based on the values measured for BPL and BT by considering the rules created in the fuzzy-model for mental stress.

Based on the feedbacks obtained through the distributed questionnaires, it was found that maximum numbers of farmers reported of stressed due to work-pressures i.e. 177 farmers (88.5%); which was followed by 146 farmers (73%) with feelings of pre-occupied, worried or tormented because work-pressures; 151 farmers (75.5%) with feeling of shoulders with great weights; 106 farmers (53%) with physical-pains: sore-back, headaches, stiff-neck, etc.; 143 farmers (71.5%) with confusion of muddled thoughts and lacking in concentration to focus on work; as well as with difficulties to control moods, emotions, reactions and/or gestures; and 128 farmers (64%) with feeling to be rushed without having enough-time to perform any tasks; respectively (Fig. 4.1). Thus, based on these findings, it was clear of the physical stresses to be resulting in mental workloads leading to mental stress among the farmers.

4.3.2 *Fuzzy Logic Modeling for Farmers' Mental Stress Based on BPL and BT*

The two inputs and one output 'mamdani-type' fuzzy-model by considering the "BPL and BT" as inputs and the "mental stress" as output, is shown in Fig. 4.2. The input-data's fuzzification was obtained by using linguistic variables. The range value's membership functions were assigned with the values as follows: Blood-Pressure Levels (BPL) in the values of '68 to 125 KPa', representing three ranges of 'Low, Medium, and High'; and Body-Temperature (BT) in the values of '24 to 31°C', representing three ranges of 'Low, Medium, and High'; respectively. Similarly, the mental stress as output was assigned in the range between '0 and 1' representing '0 = No

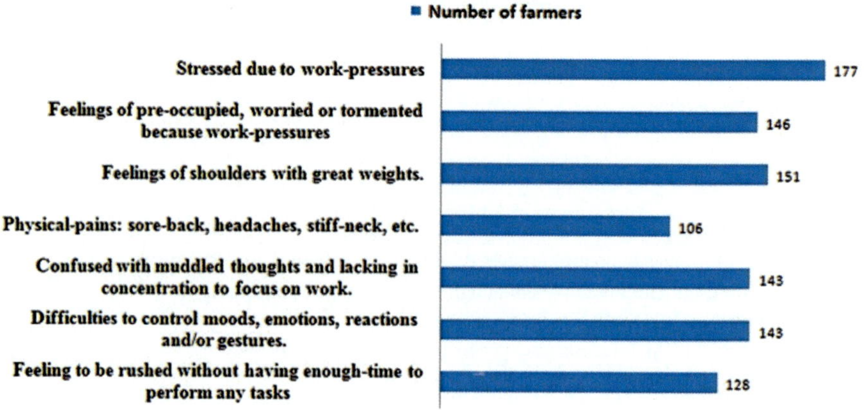

Fig. 4.1 Physical stress and mental workload among the farmers

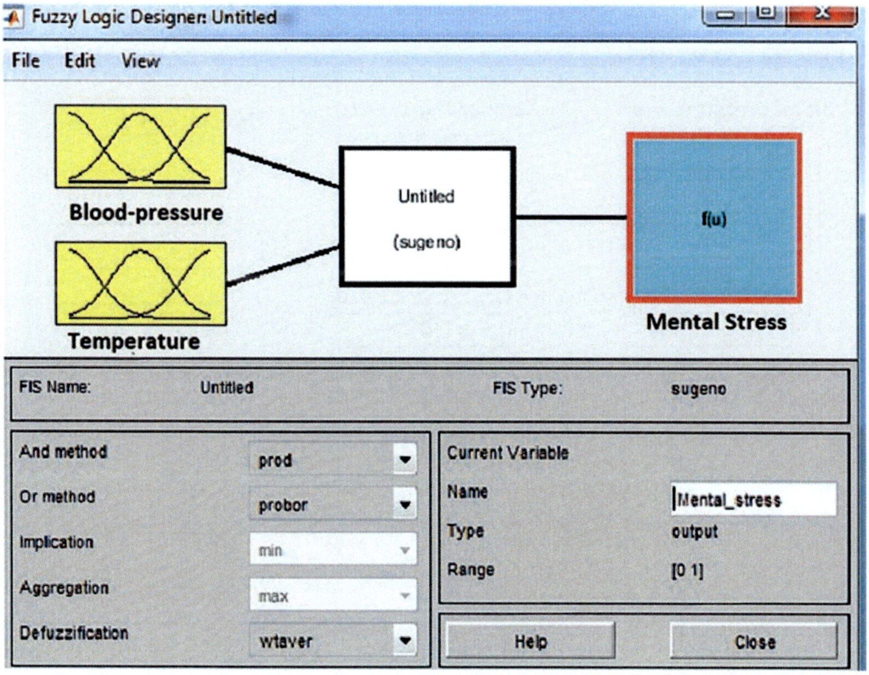

Fig. 4.2 Mamdani-type fuzzy-model with two inputs and one output

and 1 = Yes', respectively. Moreover, based on existing literature and experts' opinion, triangular membership functions were selected for all the two inputs as well as for the output.

Figure 4.3 shows the fuzzy inference system that was established in conformity to the defined rules for the model, such as: "If BPL was high and BT was high, then mental stress was yes; If BPL was high and BT was medium, then mental stress was yes; If BPL was high and BT was low, then mental stress was yes; If BPL was medium and BT was high, then mental stress was yes; If BPL was medium and BT was medium, then mental stress was no; If BPL was medium and BT was low, then mental stress was no; If BPL was low and BT was high, then mental stress was yes; If BPL was low and BT was medium, then mental stress was yes; and If BPL was low and BT was low, then mental stress was no"; respectively.

Similarly, Fig. 4.4 shows the surface that was generated for evaluating the farmers' mental stress on the basis of the considered inputs.

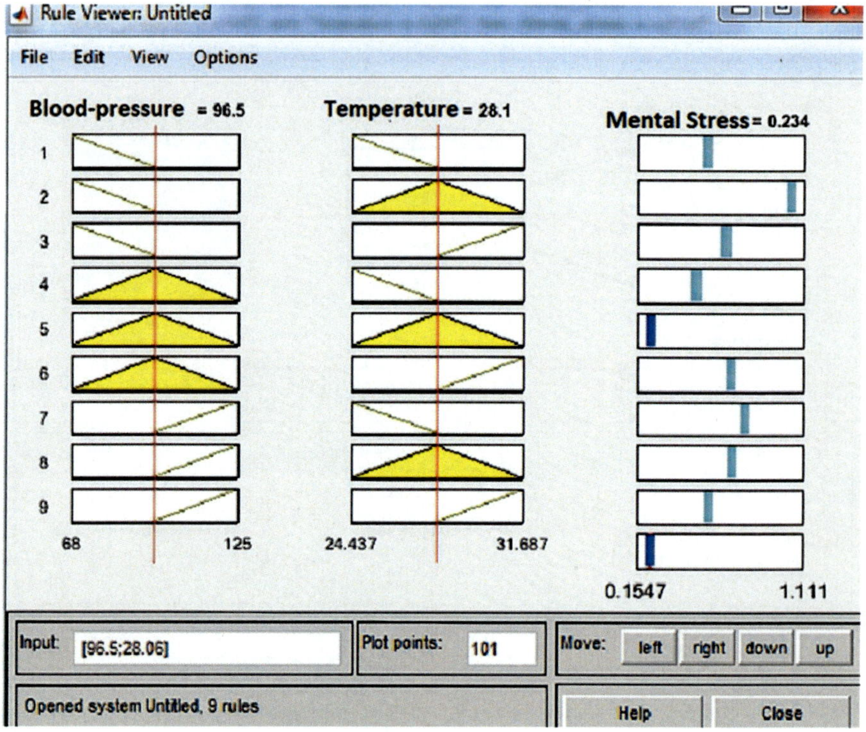

Fig. 4.3 The established model's MATLAB rules

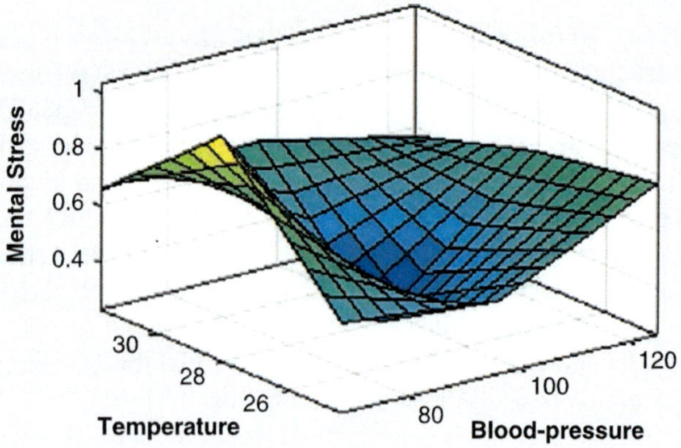

Fig. 4.4 The established model's generated surface

4.3.3 ANFIS-Based Prediction of Farmers' Mental Stress

For the generation of the ANFIS-based model, the two inputs such as BPL and BT were considered, while the farmers' mental stress was taken as output, as is illustrated in the ANFIS model-structure (Fig. 4.5).

After subsequent adjustments in the ANFIS model-structure to obtain the most effective-model, the preferred membership functions were in accordance with the developed fuzzy-model with nine rules.

The training-data with the fuzzy interface-system (FIS) output for the collected data with average testing-errors of 0.47792 is shown in Fig. 4.6a, which is further followed by the training-errors with the fifth epoch-error of 0.47792 (Fig. 4.6b), respectively. The obtained ANFIS-based information included the following: "Total-number of nodes = 35; Total-number of linear-parameters = 9; Total-number of nonlinear-parameters = 18; Total-number of parameters = 27; Total-number of 'training data-pairs' = 199; Total-number of 'checking data-pairs' = 0; Total-number of fuzzy-rules = 9; and the completion of ANFIS-training at second epoch; respectively. Moreover, it was observed that the use of ANFIS resulted in minimal-errors with good prediction-levels for farmers' mental stress.

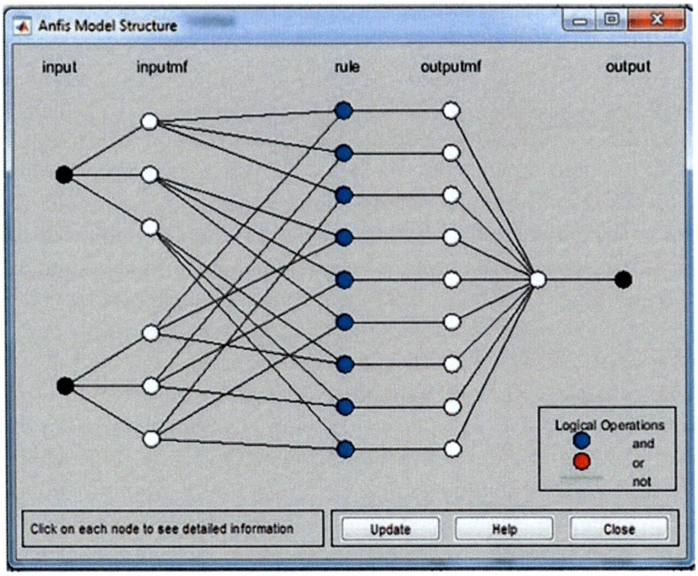

Fig. 4.5 ANFIS model-structure for farmers' mental stress

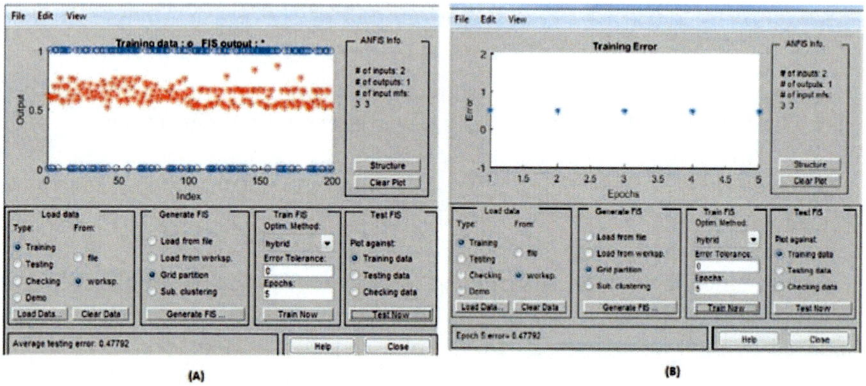

Fig. 4.6 (**a**) Training-data with the fuzzy interface-system (FIS) output for the collected data; (**b**) Associated training-error of the ANFIS-based model

4.4 Innovative Equipment to Measure Mental Workload

The present-day measurement-technology has the potential to capture the related activities that exceed the "endurance-limits of work severity" by assessing the heart-rate's reliability, the employment related key risk area, and in the quantification of the stress-situation's occurrence (Quendler et al., 2017). Several authors and researchers have considered different parameters for individual's stress-measurements. For instance, Ghaderi et al. (2015) have used "heart-rates, galvanic-skin responses (GSR), EMG and respiration", where respiration was found as a critical parameter to be associated with stresses. Similarly, Xu et al. (2015) made use of the signals like "EEG, GSR, EMG, and SpO_2" for stress measurements. Gjoreski et al. (2016) used the characteristics of both "HRV and EEG-signals" in order to predict the stress levels. In an experiment-based investigation made by Li et al. (2016), the EEG signals were recorded by utilizing the "facial-stimuli" with the use of "high-gamma as well as low-gamma rays". It was found under normal-circumstances that a higher "coherence of the brain's right-part" occurred in comparison to the patients with moderate depressed-condition. A correlation-based study was performed by Low and McCraty (2018), where they involved 85 employees working as full-time in a huge corporation with working-hours of at least 40 per week. It was found of having significant positive-correlations between "emotional-stress and heart-rate variability (HRV)" and also, between "quit-intention and HRV". Moreover, the employees with higher emotional-stress were found to be healthier along with the healthier-employees having high job quit-intentions.

Can et al. (2019) developed an "automatic stress detection-system" by using "physiological-signals" that were acquired from "unobtrusive smart-wearable-devices" with "modality-specific artefact-removal" as well as "feature-extraction methods" for real-life conditions that can be carried during daily life routine-activities of the individuals. By considering a total of 206 student's data, Ahuja and Banga

Table 4.2 Different types of IoT-based sensors used in the agricultural sectors

Sensor (Type)	Purpose	Agricultural application
Ion-selective electrode and Ion-selective field-effect transistor-sensors (Electro-Mechanical)	Detection of specific ions in soils by using electrodes	Detection of soil's Nitrogen, Potassium and Phosphorus
Tensiometer (Mechanical)	Measuring soil-compaction by using probes	Detection of the utilized forces by the roots in water-absorption
Electrodes for frequency-domain/ time-domain reflectometry (Dielectric soil Moisture)	Assessing moisture-levels by using electrodes	Measuring soil's water-content
Photo-diodes (Optical)	Measuring soil's properties by using light	Determining organic-matter, clay and moisture contents of soils
Sensor for measuring at fixed or dynamic location (Air-flow)	Measuring soil's air-permeability	For classification different soil-types, moisture-levels, and soil's structure or compaction

(2019) have tried to evaluate their stress-levels at different time-intervals. An inferring emotion-system has been proposed by Kim et al. (2020) for the use of "EEG, pulses, blood-pressures (including systolic as well as diastolic blood-pressure)", and then they recommended with "color in addition to music" for different users' emotional-states that was applicable to a user-service in "self-driving vehicle".

By taking the features from the wearable-device along with its display in mobile-application, Lakshmi et al. (2020) have proposed for an efficient solution to assess the experienced stresses of working individuals. They utilized information on "heart-rate variations, respiratory-rates, peripheral-capillary oxygen-saturation (SpO_2), and temperature-related data". Moreover, the evaluation of the system was done in a real work-environment by utilizing "fuzzy-logic algorithm", which was followed by the occupational stress-index calculation. Different machine learning algorithms were recommended by Chaware et al. (2020), such as: "Conventional Neural-Network (CNN)" in order to extract the 'Facebook-posts', "Transductive-Support Vector-Machine (TSVM)" for the post's classification, and "K-Nearest-Neighbors (KNN)" in order to recommend the nearby-hospitals. In their study the stress-level was predicted as "positive or negative" for persons based on such algorithms. Several IoT-based sensors are available for various agriculture purposes (Boursianis et al., 2020). Table 4.2 illustrates different types of IoT-based sensors used in the global agricultural sectors.

4.5 Conclusion

Any individual's physical as well as mental load are associated to the stress for which the organisms try in adapting the use of some kind of defence-mechanisms. Although the prevailing stresses may be considered as part of life, but in-turn these stresses can directly and/or indirectly result in the occurrence of different types of unwanted and intolerable diseases with its continuation for longer-periods.

Moreover, the prevalence of mental stress among individuals can affect their performance levels when trying for wide range of complex tasks by them. Indeed, exposing to higher levels of mental stress is usually not tolerable that can affect the successful completion of any tasks. For the evaluation of the work-related loads and induced mental stress, the measurement of associated variables can be very useful for all levels for possible intervention-strategies including the decision as well as policy makers.

4.5.1 Recommendations with Proposed Equipment or Devices

The emotional or mental state is regarded as stress that results in tension-creation due to any adverse-circumstances. The individuals nowadays are found in facing of stresses with varying levels for different people. In the recent times, assessing the human-brain signals through the use of EEG-technology has been proved to be an innovative idea to evaluate the individual's stress-levels. This can be achieved by mobile-phones with suitable applications for capturing the recorded EEG-signals from EEG-amplifiers, and then with subsequent analysis in "LabView" for the assessment of mental stress levels.

Moreover, although a number of standard devices are available commercially for measuring different parameters that are closely associated with the mental condition of individuals, but measuring each individual parameter by the use of those devices seem to be time-consuming and also, carrying of individual devices are not suitable in most of the cases. Thus, by integrating the individual devices into a single one can be more beneficial to the users.

By the use of Raspberry pi and different sensors for measuring the parameters can provide more accurate and quicker outcomes as compared to the conventional ones. The sensors can be assembled with the Raspberry pi and the output can be obtained by the use of any display monitor or mobile phone for better interpretation of the obtained data (Kumar & Srinivasan, 2020). For instance, in view of the parameters that was measured in this study by the use of standard devices included: the "Blood-pressure levels and Body-temperature" which can be measured by using any wearable-device (Can et al., 2019) coupled with Raspberry pi and display monitor or mobile-phone to obtain the proper correlated state of mind or the mental stress (Fig. 4.7). However, the number of input measuring parameters can be increased based on the cases and requirements for more accurate evaluation of mental stress (Fig. 4.8).

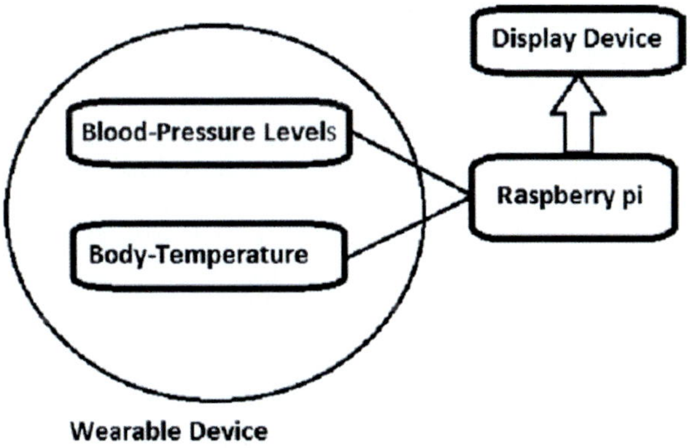

Fig. 4.7 Proposed device to measure the mental stress associated parameters

Fig. 4.8 measuring device

References

Ahuja, R., & Banga, A. (2019). Mental stress detection in university students using machine learning algorithms. International conference on pervasive computing advances and applications–PerCAA 2019. *Procedia Computer Science, 152*, 349–353.

Akay, D., Chen, X., Barnes, C., & Henson, B. (2012). ANFIS modeling for predicting affective responses to tactile textures. *Human Factors and Ergonomics in Manufacturing & Service Industries, 22*(3), 269–281.

Alberdi, A., Aztiria, A., & Basarab, A. (2016). Towards an automatic early stress recognition system for office environments based on multimodal measurements: A review. *Journal of Biomedical Informatics, 59*, 49–75.

Al-Hmouz, A., Shen, J., Al-Hmouz, R., & Yan, J. (2011). Modeling and simulation of an adaptive neuro-fuzzy inference system (ANFIS) for mobile learning. *IEEE Transactions on Learning Technologies, 5*(3), 226–237.

American Psychology Association. (2019). *Stress: The different kinds of stress* (p. 2019). American Psychology Association.

Bahram-Parvar, M., Salehi, F., & Razavi, S. M. (2017). Adaptive neuro-fuzzy inference system (ANFIS) simulation for predicting overall acceptability of ice cream. *Engineering in Agriculture, Environment and Food, 10*(2), 79–86.

Berkowitz, A. D., & Perkins, H. W. (1984). Stress among farm women: Work and family as interacting systems. *Journal of Marriage and the Family*, 161–166.

Boursianis, A. D., Papadopoulou, M. S., Diamantoulakis, P., Liopa-Tsakalidi, A., Barouchas, P., Salahas, G., Karagiannidis, G., Wan, S., & Goudos, S. K. (2020). Internet of things (IoT) and agricultural unmanned aerial vehicles (UAVs) in smart farming: A comprehensive review. *Internet of Things*. https://doi.org/10.1016/j.iot.2020.100187

Brigance, C., Soto Mas, F., Sanchez, V., & Handal, A. J. (2018). The mental health of the organic farmer: Psychosocial and contextual actors. *Workplace Health & Safety, 66*(12), 606–616.

Can, Y. S., Chalabianloo, N., & Ersoy, D. E. C. (2019). Continuous stress detection using wearable sensors in real life: Algorithmic programming contest case study. *Sensors, 19*, 1849. https://doi.org/10.3390/s19081849

Cezar-Vaz, M. R., Bonow, C. A., & Da Silva, M. R. S. (2015). Mental and physical symptoms of female rural workers: Relation between household and rural work. *International Journal of Environmental Research and Public Health, 12*(9), 11037–11049.

Chakraborty, N., & Barma, M. D. (2014). Modelling of stand-alone wind energy conversion system using fuzzy logic controller. *Small, 2*(1).

Chaware, S. M., Makashir, C., Athavale, C., Athavale, M., & Baraskar, T. (2020). Stress detection methodology based on social media network: A proposed design. International journal of innovative technology and exploring. *Engineering, 9*(3). https://doi.org/10.35940/ijitee.B7537.019320

Colligan, T. W., & Higgins, E. M. (2006). Workplace stress: Etiology and consequences. *Journal of Workplace Behavioral Health, 21*, 89–97.

Daghagh Yazd, S., Wheeler, S. A., & Zuo, A. (2019). Key risk factors affecting farmers' mental health: A systematic review. *International Journal of Environmental Research and Public Health, 16*(23), 4849.

de Santos, S. A., Ávila, C. S., Casanova, J. G., & del Pozo, G. B. (2011a). A stress-detection system based on physiological signals and fuzzy logic. *IEEE Transactions on Industrial Electronics, 58*(10), 4857–4865.

de Santos, S. A., Ávila, C. S., del Pozo, G. B., & Casanova, J. G. (2011b). Stress detection by means of stress physiological template. In *2011 third world congress on nature and biologically inspired computing* (pp. 131–136). IEEE.

El-Samahy, E., Mahfouf, M., Torres-Salomao, L. A., & Anzurez-Marin, J. (2015). A new computer control system for mental stress management using fuzzy logic. In *2015 IEEE international conference on evolving and adaptive intelligent systems (EAIS)* (pp. 1–7). IEEE.

England, M. J., Liverman, C. T., Schultz, A. M., & Strawbridge, L. M. (2012). Epilepsy across the spectrum: Promotinghealth and understanding: A summary of the Institute of Medicine report. *Epilepsy & Behavior, 25*, 266–276.

Fragiadakis, N. G., Tsoukalas, V. D., & Papazoglou, V. J. (2014). An adaptive neuro-fuzzy inference system (anfis) model for assessing occupational risk in the shipbuilding industry. *Safety Science, 63*, 226–235.

Furey, E. M., O'Hora, D., McNamara, J., Kinsella, S., & Noone, C. (2016). The roles of financial threat, social support, work stress, and mental distress in dairy farmers' expectations of injury. *Frontiers in Public Health, 4*, 126.

Gajjar, S., Sarkar, M., & Dasgupta, K. (2014). Cluster head selection protocol using fuzzy logic for wireless sensor networks. *International Journal of Computer Applications, 97*(7).

Ghaderi, A., Frounchi, J., & Farnam, A. (2015). Machine learning-based signal processing using physiological signals for stress detection. In *22nd Iranian conference on biomedical engineering (ICBME)-2015* (pp. 93–98).

Ghatul, D. B. (2013). Constraints and stress level of farmers. *International Journal of Scientific and Research Publications, 107*.

Gjoreski, M., Gjoreski, H., Lustrek, M., & Gams, M. (2016). Continuous stress detection using a wrist device: In the laboratory and real life. In *Proceedings of the 2016 ACM international joint conference on pervasive and ubiquitous computing: adjunct* (pp. 1185–1193). ACM.

Gregoire, A. (2002). The mental health of farmers. *Occupational Medicine, 52*(8), 471–476.

Gulhane, Y. H., Rode, S. V., & Ladhake, S. A. (2011). Application of fuzzy logic in stress analysis. In *Proceedings of the international conference & workshop on emerging trends in technology* (pp. 679–685).

Hagen, B. N., Albright, A., Sargeant, J., Winder, C. B., Harper, S. L., O'Sullivan, T. L., & Jones-Bitton, A. (2019). Research trends in farmers' mental health: A scoping review of mental health outcomes and interventions among farming populations worldwide. *PLoS One, 14*(12), e0225661.

Hovey, J. D., & Seligman, L. D. (2006). The mental health of agricultural workers. In *Agricultural medicine* (pp. 282–299). Springer.

Imbeau, D., Desjardins, L., Dessureault, P. C., Riel, P., & Fraser, R. (1995). Oxygen consumption during scaffold assembling and disassembling work: Comparison between field measurements and estimation from heart rate. *International Journal of Industrial Ergonomics, 15*, 247–259.

Kallioniemi, M. K., Simola, A., Kaseva, J., & Kymäläinen, H. R. (2016). Stress and burnout among Finnish dairy farmers. *Journal of Agromedicine, 21*(3), 259–268.

Kearney, G. D., Rafferty, A. P., Hendricks, L. R., Allen, D. L., & Tutor-Marcom, R. (2014). A cross-sectional study of stressors among farmers in eastern North Carolina. *North Carolina Medical Journal, 75*(6), 384–392.

Kim, T.-Y., Hoon, K. O., & Kim, S.-H. (2020). Data analysis for emotion classification based on bio-information in self-driving vehicles. *Journal of Advanced Transportation, 2020*, 11. https://doi.org/10.1155/2020/8167295

Koochaksaraei, A. A., & Izadfar, H. (2019). High-efficiency MPPT controller using ANFIS-reference model for solar systems. In *2019 5th Conference on knowledge based engineering and innovation (KBEI)* (pp. 770–775). IEEE.

Kumar, P. A., & Srinivasan, P. (2020). Intergration for Mearsuring blood pressure and body temperature sensors using Mobile application. NCICCT – 2020 conference proceedings. *International Journal of Engineering Research & Technology, 8*(8).

Kushwaha, G. S., & Kumar, S. (2009). Role of the fuzzy system in psychological research. *Europe's Journal of Psychology, 5*(2), 123–134.

Lakshmi, B., Boban, E., Sulphikar, N., & Aparna, T. (2020). Mental stress calculation using fuzzy logic algorithm. *International Journal of Applied Engineering Research, 15*(1).

Lazaridis, E. N. (1999). A Bayesian evaluation of fuzzy logic in a classification problem. *Stochastic Models, 15*(3), 461–483.

Lemyre, L., Tessier, R., & Fillion, L. (1990). *La Mesure du stress psychologique: manuel d'utilisation*. Behaviora.

Li, X., Jing, Z., Hu, B. (2016). *An EEG-based study on coherence and brain networks in mild depression cognitive process*. Shuting Sun School of Information Science & Engineering, Lanzhou University Lanzhou, China, Conference Paper, December-2016. https://doi.org/10.1109/BIBM.2016.7822702.

Low, A., & McCraty, R. (2018). Heart rate variability: New perspectives on assessment of stress and health risk at the workplace. *Heart Mind, 2*, 16–27.

Mallikarjun, H. M., Akshay, C., Apoorva, G. S., Gowri, J., & Sheetal, B. V. (2019). Depression predictor model for farmers using machine learning techniques. *International Journal of Innovative Technology and Exploring Engineering, 9*(2S). https://doi.org/10.35940/ijitee.B1047.1292S19

MATLAB. (2013). The MathWorks, Inc., Natick, Massachusetts, United States.

Milczarek, M., & Elke Schneider, E. G. (2009). *OSH in figures, stress at work, fact and figures*. European Agency for Safety and Health at Work.

Mishra, D., & Satapathy, S. (2020). Application of fuzzy expert system for prediction of farmer muscle strength: A collective database and analysis in agricultural sectors of Odisha in India. In A. Kumar & M. Kalpana (Eds.), *Fuzzy expert systems and applications in agricultural diagnosis* (pp. 195–238). IGI Global. https://doi.org/10.4018/978-1-5225-9175-7.ch011

Mittal, S., & Grewal, N. S. (2014). Stress quantification using fuzzy analysis of ECG parameters. *International Journal of Computer Applications, 99*(9).

Okwu, M. O., & Adetunji, O. (2018). A comparative study of artificial neural network (ANN) and adaptive neuro-fuzzy inference system (ANFIS) models in distribution system with nondeterministic inputs. *International Journal of Engineering Business Management, 10*.

Padhy, C., & Raju, P. S. (2018). Stress among farmers and its alleviation. *International Journal of Management, Technology And Engineering, 8*(12), 2882–2887.

Pandey, A., & Parhi, D. R. (2017). Optimum path planning of mobile robot in unknown static and dynamic environments using fuzzy-wind driven optimization algorithm. *Defence Technology, 13*(1), 47–58.

Picard, R. W. (2016). Automating the recognition of stress and emotion: From lab to real-world impact. *IEEE Multimedia, 23*, 3–7.

Quendler, E., Trieb, K., & Nimmerichter, A. (2017). Validation of automated detection of physical and mental stress during work in a Hühnermobil 225. *Annals of Agricultural and Environmental Medicine, 24*(2), 329–331. https://doi.org/10.5604/12321966.1235183

Ramanathan, P. (2014). Fuzzy logic controller for temperature regulation process. *Middle-East Journal of Scientific Research, 20*(11), 1524–1528.

Rammelmeier, T., Weisner, K., Günthner, W. A., Deuse, J. (2014). Reduktion der Mitarbeiterbelastung in der Kommissionierung auf Basis einer fortlaufenden Belastungsermittlung. (Hrsg.): Gesellschaft für Arbeitswissenschaft e.V., Gestaltung der Arbeitswelt der Zukunft, GfA–Press, München, 2014.

Şahin, M., & Erol, R. (2017). A comparative study of neural networks and ANFIS for forecasting attendance rate of soccer games. *Mathematical and Computational Applications, 22*(4), 43.

Sawhney, A., Bhatia, R., & Mahajan, P. (2014). Congestion control in wireless communication network using fuzzy logic and machine learning techniques. *International Journal of Advanced Research in Electrical, Electronics and Instrumentation Engineering, 3*(11).

Shatte, A. B. R., Hutchinson, D. M., & Teague, S. J. (2019). Machine learning in mental health: A scoping review of methods and applications. *Psychological Medicine, 1-23*. https://doi.org/10.1017/S0033291719000151

Taouche, R. (2019). Fuzzy logic prediction of stress-strain behavior parameters of two-phase materials. Application to WC-Co system. *Australian Journal of Basic and Applied Sciences, 13*(6), 74–81.

Tiwari, S., Babbar, R., & Kaur, G. (2018). Performance evaluation of two ANFIS models for predicting water quality index of River Satluj (India). *Advances in Civil Engineering, 2018*.

Torres-Salomao, L. A., Mahfouf, M., & Obajemu, O. (2014). Interval type-2 fuzzy logic adaptive modelling for human operators undergoing mental stress. *IFAC Proceedings, 47*(3), 9880–9885.

Walia, N., Singh, H., & Sharma, A. (2015). ANFIS: Adaptive neuro-fuzzy inference system-a survey. *International Journal of Computer Applications, 123*(13).

Xu, Q., Nwe, T. L., & Guan, C. (2015). Cluster-based analysis for personalized stress evaluation using physiological signals. *IEEE Journal of Biomedical and Health Informatics, 19*(1), 275–281.

Yadollahpour, A., Nourozi, J., Mirbagheri, S. A., Simancas-Acevedo, E., & Trejo-Macotela, F. R. (2018). Designing and implementing an ANFIS based medical decision support system to predict chronic kidney disease progression. *Frontiers in Physiology*, 9.

Chapter 5
Personal Protective Equipment for Farmers

5.1 Introduction

Occupational health and safety (OHS) has been a first-rate challenge globally spreading from commercial to agriculture sectors. A range of hazards having undesirable consequences are associated in farming like infection by pathogens, injuries from exposure, physiological disorders, poisoning, metabolism infections, and musculoskeletal disorders, etc. According to data from the "Centers for Disease Control and Prevention" (CDC, 2013), agriculture is one of the three most dangerous occupations in terms of work-related injuries and diseases. This can be achieved in the agricultural sector, with special attention to health and safety issues in this sector (Jugmohan, 2013). The agricultural workers have to work in hazardous condition like exposure to sunlight, dust, dirt etc. while performing different activities like weeding, harvesting, threshing, de-husking, cleaning and packaging of crops. Many of them were not aware about the harmful effect created by strong UV light of the Sun and other microorganisms mixed with soil and also dust particles.

It is only protective clothing which protects individuals who are exposed to life threatening or hazardous environment during work. Today, the use of data-analytics, automations, connectivity, and remote-tracking have made great strides, bringing innovation to all fields of modern civilization including smart-farming (Alahi et al., 2017; Balaji et al., 2019; Ibarra-Esquer et al., 2017; Köksal & Tekinerdogan, 2019; Rains & Thomas, 2009; Sørensen et al., 2011; Turner et al., 2000; Villa-Henriksen et al., 2020). The Internet of Things (IoT) has revolutionized the digitization of daily-life. Early detection or easier detection of plant diseases or health conditions will enable the farmers to develop appropriate prevention strategies. There are many examples of how the IoT makes human activities easier. All types of IoT-based devices can collect and exchange information from their environment. An interconnected world will bring potentially unlimited possibilities for most types of human activities.

S. Satapathy et al., *Innovation in Agriculture with IoT and AI*,
Innovation, Technology, and Knowledge Management,
https://doi.org/10.1007/978-3-030-88828-2_5

The farmers are subjected to various type of injuries, when adequate safety measures are not under taken. Moreover, it was observed that during the weeding activities, when the conventional sickles or knives are used, then it may lead to different unavoidable injuries to the operators. Thus, it becomes essential for protecting these farmers from agriculture related accidents or injuries as well as from adverse climatic changes, to make them comfortable and productive while performing the agricultural activities.

5.2 Farm Related Injuries and the Usefulness of Personal Protective Equipment

Abrahão et al. (2012) studied the use of personal protective equipment, especially the use of protective gloves when manually cutting sugar cane, and found the problem of neglecting the test standards in the manufacturing process of personal protective equipment. DeMuri and Purschwitz (2000) emphasized that violations in the agricultural sector are caused by lack of supervision, unreasonable expectations, economic problems and insufficient safety precautions. Xiang et al. (2000) conducted personal interviews with 1358 Chinese farmers out of 1500 interviewed farmers from 14 villages to assess agricultural damage. It showed higher death and injury rates, from 0.22 to 166 per 1000 workers per year. In the American Agricultural Survey (Rautiainen & Reynolds, 2002). In Madhya Pradesh, India, the agricultural accident rate was 1.25 per thousand workers per year. It was also found that 77.6% of accidents were caused by farms and machines, while 11.8% were related to the use of hand tools and other reasons (Tiwari et al., 2002). According to reports, the number of injured farmers was higher than that of industrial workers (Helkamp & Lundstrom, 2002).

Kumar et al. (2008) reported a total of 576 agricultural-related injuries when using appropriate hand tools, which was 332 or 58%, where 19,723 people in 9 villages in the first phase were taken into account. Additionally, in second phase of study by considering more 21 villages with 78,890 persons, 54 i.e.19% of hand tool related injuries out of 282 injuries was reported. It was further recommended to have intervention training and development activities at block-levels for the safety measures of tools and equipment. In an investigation, the agricultural accidents rate was reported as 6.39 per thousand workers per year with farm-implement related injury of 40% (Kumar & Dewangan, 2009). In another study carried out in Odisha (India), it was observed that out of a total of 45 farmers, 43 farmers were victims of farm related accidents due to the use of hand tools (Mishra & Satapathy, 2018a). Dube et al. (2018) have made an ergonomic assessment for different postures of farmers of Odisha in India during seeding, fertilizing and weeding processes in agricultural field, and based on the obtained RULA scores, different action levels were suggested for the existing working techniques and methods.

5.2.1 Use of Hand Gloves in Agriculture

Agricultural pesticides can be hazardous if not used safely. Under typical working conditions, the majority of exposure to the agricultural pesticides is through the hand-skin. Wearing appropriate gloves should be a standard practice when handling these. In a survey conducted across the Slovak Republic, about the practices of the use of protective gloves it was found that 75% of respondents' wear gloves when handling the agricultural pesticides. While, 58% use the same disposable gloves repeatedly, 67% wear the reusable gloves from last season, and 53% do not wash outside of the reusable gloves with detergent before removing them. 30% of glove users wear all available gloves including absolutely inappropriate materials like fabric and leather, 28% do not recognize the glove material which they use, 20% use disposable Latex and 3% Vinyl and Polyethylene gloves. Only 19% of glove users wear a material that is generally considered a good choice for handling the agricultural pesticides (Hojerová et al., 2015).

There are many gloves that can deal with different types of hazards in the work environment, but no single glove can guarantee protection against all hazards. According to the intended use, there are medical gloves, household cleaning gloves, food processing gloves, gardening gloves, etc. Gloves for handling synthetic materials and liquids and gloves for special purposes. Manufacturers use different standards to organize gloves, such as material, thickness, texture and structure, weight, chemical composition, heat and cold protection, mechanical protection, water and chemical protection (Fishel, 2015). Nelson and Mital (1995) made gloves of different thicknesses at home, limiting other installation options and finding a reasonable balance between thickness and puncture resistance. Tremblay-Lutter and Weihrer (1996) found in hand measurements that negative slack (tight fit) along the length and circumference of the fingers, but positive slack around the palm circumference (loose fit) is the best. Fisher et al. (1999) concluded that the use of double-layer gloves reduces the possibility of frustration, but there are conflicting results in the effect of thickness. Kwon et al. (2009) determined the relationship between arm measurements to decide the critical dimensions for evaluation, and recommended that arm length and circumference are the most representative; they also specified different sizing systems for men and women. In addition, a total of 36 agricultural injuries were reported in 12 villages in Punjab, India in one year. Furthermore, the annual injury rates because of the use of tractors, sprinklers, electric motors, threshers, and feed harvesters per 1000 vehicles were 23.7, 15.5, 7.1, 5.7, and 2.2, respectively (Mittal et al., 1996). Mishra and Satapathy (2018b) have studied the musculoskeletal disorders during threshing of crops for farmers in Odisha, and suggested to prefer ergonomic design of farming tools.

5.2.2 Need of Protective Cap for Farmers

Throughout the world, several studies have been made that examined the potential-effects of climate change on agriculture (Adams et al., 1990; Adams et al., 1995; Mendelsohn et al., 1994; Kurukulasuriya & Ajwad, 2007; Mariara & Karanja, 2007). It becomes very much important for adaptation, if the farmers are encountering or about to face the potential unfavorable climate change impacts (Mariara & Karanja, 2007; Nhemachena & Hassan, 2007). Further, it will help in protecting the livelihoods of poor-farmers, and will also ensure in food-security with considerable reduction in the potential negative-impacts and with subsequent reinforcements of the associated advantages with climate changes (Bradshaw et al., 2004; IPCC, 2001; Reid et al., 2007).

In a study conducted by Gogoi et al. (2016) in "Jorhat district of Assam (India)" with the aim to obtain the occupational health-hazards of worker's engaged in various agricultural activities, they made an attempt to study the consciousness regarding the dressing patterns to combat the hazardous working-condition. The primary-data were collected with the help of questionnaire, and a protective clothing-kit was developed. Further, the acceptability of developed garments was also analyzed. Four central-components related to agriculture were mainly examined by Humayun Kabir (2015), such as: "The rate of self-reported acceptance of adaptive instruments (coping-plans) based on the consequences of climate changes; Ranking of the potential coping-plans based on their apparent significance to agricultural-initiatives; Documentation of the socio-economic factors related with adoption of coping-plans, and Ranking of the potential limitations to adoption of coping-plans based on farmers' reporting on the restrictions", respectively.

5.3 Recommended Design Concepts for the Global Farming-Community

5.3.1 Design of Multi-Purpose Hand Gloves

The hand gloves need to be cut resistive to protect the hand of farmers. Thus, the proposed hand glove design consisted of a meshing of steel materials, which make it cut resistive. As shown in Fig. 5.1, a saw type cutter can be connected to the glove material having six projected saws. The mid-section saws i.e. third and fourth saws can be pivoted together, so as to make it convenient for the first finger to bend as required during cutting. The hand gloves need to be made of meshed steel materials that will prevent the hands like palms and fingers from external injuries during weeding operations, in particular. The cutting of the unwanted plants can be carried out by the sawing action of the first finger of the hand gloves. Thus, it eliminates the external injuries from the use of sickles. Moreover, IoT-based features can be

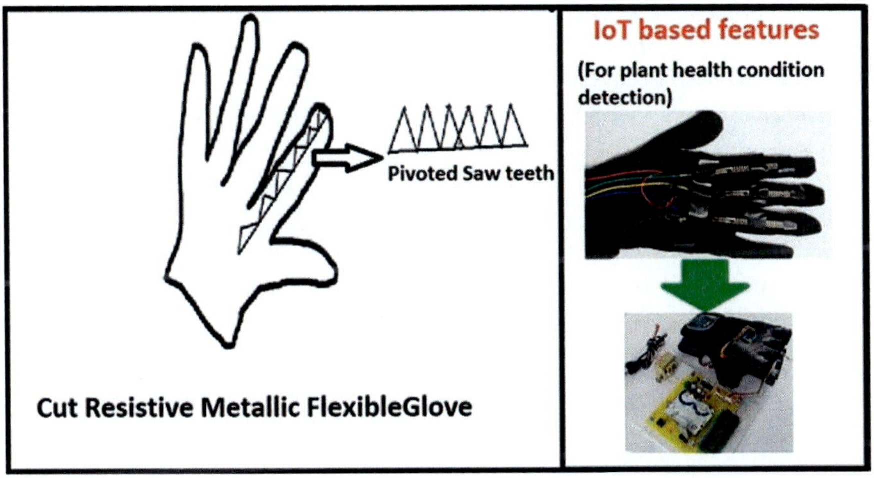

Fig. 5.1 Proposed model for farmer's hand gloves

additionally added in the gloves to detect different plant diseases or condition of plants with the help of attached sensors.

5.3.1.1 Benefits to the Society

In the cutting edge cultivation activities in agriculture, the utilization of defensive and security gadgets during completing the farm work significantly expands the farmer's laborer efficiency. In current circumstances, further developed strategies have supplanted many farming occupations in the past did by difficult work or by the help of draft animals. The whole history of agriculture contains numerous instances of the utilization of devices, like the tool and the furrow. Automation was one of the huge component answerable for urbanization and mechanical frugalities. Other than further enhancements in proficiency, automation empowers enormous scope creation. However, it can dislodge incompetent homestead work and can cause natural exploitations (like contamination, deforestation, and soil disintegration), particularly on the off chance that it is applied limitedly instead of comprehensively. It builds proficiency and per man usefulness.

Mechanization raises the productivity of work and upgrades the farmers' outcomes per laborer. By its effective utilization, it can help to decrease the quantum of work expected to deliver a unit of yield. Early developments were based on execution and utilizations of apparatuses that expanded the efficiency of draft animals and helped the farmers in getting ready land for development, planting and cultivating, and collecting crops. In addition, mechanization and automations ought not be presented in a rush, or on a too huge scope. To be fruitful, it ought to be steadily extended and kept inside capability norms of the individuals who work with it. Subsequently, the proposed hand gloves can be a less expensive kind of weed controlling hand glove, which will be reasonable for various environment conditions, and would assist the worker or farmer with performing task more productively instead of utilizing ordinary techniques like sickles.

The benefits of proposed design in agriculture sectors include:

- It will result in lower cost of work.
- It contracts the demand for conventional sickles for weeding operations.
- It will bring in improvements in agricultural technique by a more scientific approach.
- It will modify social structure in rural areas.
- It will lead to commercial and safe agriculture.
- More versatile to operate.
- More durable, and easy to operate.

Moreover, the advantages of the proposed hand gloves over the existing technologies include the following:

- Helps to cut unwanted plants in weeding without the use of sickles.
- Simple to operate manually.
- The steel materials enhance the gloves durability.
- The design is based on anthropometric dimensions of hand.
- Protect the hands from external injuries.
- Reduce the drudgery in operation.
- Useful in the detection of plant health conditions with the aid of IoT-based features.

5.3.2 *Design of Multi-Purpose Farmers' Protective Hats*

The proposed design shown in Fig. 5.2 is a protective cap for the farmers who work in all climatic conditions in the agricultural sectors. This cap can be designed in an advanced and modernized way according to the need of farmers to reduces their stresses, so that they can give more effort on their working and also, will be able to enhance their productivity-levels. The specialization of the cap includes its multi-purpose applications providing more comfort and utilization in all seasons. This cap can be made as hollow form at upper side, which can be attached with solar panels to obtain the required power for different power-based appliances. In font side of the cap, a flash LED light can be attached, so that the farmer can also work easily at evening-time. With the attachment of an Alexa, direct communications with satellite can be made by the farmers regarding any needed suggestion or information. Alexa is also capable for entertaining purposes, such as music and radio etc. With the attachment of fiber glass at the front side, adequate protections can be provided to the farmers from dusts and other elements. However, by providing some portable fan behind the cap as an attachment, the farmer can be able to get cool air and

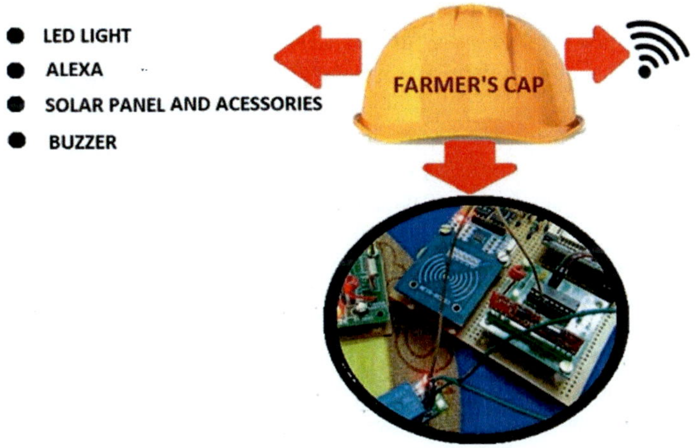

Fig. 5.2 Proposed model for farmer's protective hat

becomes comfortable in working with hot climatic conditions. All these devices can be attached with solar panel.

The major advantages of the proposed protective cap include the following:

- Can be made eco-friendly.
- Capable of handling and carrying in adverse climate conditions.
- Can be useful to improve the overall productivity of farming.
- Can be useful to reduce the mental stresses in farming.
- Can be useful to enable the small-scale farmers to implement latest technology in farming.
- Can be useful to protect the farmers from all climatic condition.
- Easy to wear.
- Light in weight.
- Useful to perform work activities in evening time.
- Can be useful to connect with satellite for appropriate information retrieval.

5.4 Conclusion

The IoT technology has been pushing the agricultural enterprises into a new revolution. By providing information about soil-moisture, chemical's use, animal health, etc., adequate preventive maintenance strategies can be under taken in the agricultural sectors. The IoT-based agriculture helps farmers around the world to not only increase productivity-levels, but also to maximize their resources. The use of IoT technology to develop agriculture will help farmers to keep production in line with the ever-increasing demand for food.

In addition, the methods used in traditional agriculture systems require greater manual inputs, physical presence, and experience in pruning plants from seed to harvest using manual tools. The use of hand gloves with additional cutter function can simplify plant cutting operations with the use of hand tools, and will also protect farmers from related injuries. In recent years, smart sensors in agriculture have attracted widespread attention. The use of IoT-based tools and techniques in agriculture will help in planning of various activities by making the farm-related task's execution more successful with limited resources and with minimal human-intervention. Therefore, IoT-based hand gloves and other personal protective equipment can increase the agricultural efficiency and productivity of the global agricultural communities.

The smart personal protective equipment has been available for several years, and its use can find new ways to deal with current world events, as they work through connections between smart wearable devices, biometric-technology, and centralized cloud-networks. Moreover, the personal protective equipment using wearable technology with embedded electronic-devices, such as tracking devices, sensors as well as monitors, can help in providing information about farmers' and plants' health-conditions and possible harm to or by the surrounding environment.

References

Abrahão, R. F., Gonzaga, M. C., & Braunbeck, O. A. (2012). Protective gloves on manual sugar cane cutting are really effective? *Work, 41*, 4963–4966. https://doi.org/10.3233/WOR-2012-0039-4963

Adams, R. M., Rosenzweig, C., Peart, R. M., Ritchie, J. T., McCarl, B. A., et al. (1990). Global climate change and US agriculture. *Nature, 345*, 219–224.

Adams, R. M., Fleming, R. A., Chang, C. C., McCarl, B. A., & Rosenzweig, C. (1995). A reassessment of the economic effects of global climate change on US agriculture. *Climatic Change, 30*, 147–167.

Alahi, M. E. E., Li, X., Mukhopadhyay, S., & Burkitt, L. (2017). A temperature compensated smart nitrate-sensor for agricultural industry. *IEEE Transactions on Industrial Electronics, 64*(9), 7333–7341.

Balaji, S., Nathani, K., & Santhakumar, R. (2019). IoT technology, applications and challenges: A contemporary survey. *Wireless Personal Communications, 108*(1), 363–388.

Bradshaw, B., Dolan, H., & Smit, B. (2004). Farm-level adaptation to climatic variability and change: Crop diversification in the Canadian prairies. *Climatic Change, 67*, 119–141.

CDC. (2013). *Agricultural safety*. Centers for Disease Control and Prevention. Retrieved from www.cdc.gov/niosh/topics/aginjury/

DeMuri, G. P., & Purschwitz, M. A. (2000). Farm injuries in children: A review. *Wisconsin Medical Journal, 99*(9), 51–55.

Dube, B., Mishra, D., Satapathy, S., & Singh, S. (2018). Ergonomic evaluation of seeding, fertilizing and weeding postures in agricultural field. *International Journal of Pure and Applied Mathematics, 119*(10 Special Issue B), 1875–1880.

Fishel, F. M. (2015). *Glove selection for working with pesticides*. UF/IFAS Extension. PI-120. Available in: http://edis.ifas.ufl.edu/.

Fisher, M. D., Reddy, V. R., Williams, F. M., Lin, K. Y., Thacker, J. G., & Edlich, R. F. (1999). Biomechanical performance of latex and non-latex double-glove systems. *Journal of Biomedical Materials Research, 48*(6), 797–806.

Gogoi, N., Choudhury, S., & Gogoi, M. (2016). Designing and construction of protective clothing for agricultural workers. *International Journal of Engineering and Technical Research, 6*(3), 2454–4698.

Helkamp, J., & Lundstrom, W. (2002). Tractor related deaths among West Virginia farmers. *Annals of Epidemiology, 12*(7), 510. https://doi.org/10.1016/S1047-2797(02)00344-7

Hojerová, J., Beránková, M., Peráčková (neé Klimová) Z, & Birbičová, S. (2015). Protective gloves use and work habits of non-professionals handling agricultural pesticides-a survey. *Acta Chimica Slovaca, 8*(2), 133–139. https://doi.org/10.1515/acs-2015-0023

Humayun Kabir, M. D. (2015). Impact of climate change on agriculture especially in Jessore and Sathkhira districts according to farmers' mitigation strategies to climate change; evidence from farmer level data. *Journal of Geography and Natural Disasters, 5*(3). https://doi.org/10.4172/2167-0587.1000152

Ibarra-Esquer, J., González-Navarro, F., Flores-Rios, B., Burtseva, L., & Astorga-Vargas, M. (2017). Tracking the evolution of the internet of things concept across different application domains. *Sensors, 17*, 1379.

IPCC. (2001). *Climate change 2001: Impacts, adaptation, and vulnerability: Contribution of working group II to the third assessment report of the intergovernmental panel on climate change*. Cambridge University Press.

Jugmohan, R. (2013). *Workshop for the Caribbean on the FAO/UNFPA guidelines: Integrating population and housing with agricultural censuses; with selected country practices*. Ministry of Planning and Sustainable Development.

Köksal, Ö., & Tekinerdogan, B. (2019). Architecture design approach for IoT-based farm management information systems. *Precision Agriculture, 20*, 926–958. https://doi.org/10.1007/s11119-018-09624-8

Kumar, G. V. P., & Dewangan, K. N. (2009). Agricultural accidents in north eastern region of India. *Safety Science, 47*, 199–205. https://doi.org/10.1016/j.ssci.2008.03.007

Kumar, A., Singh, J. K., Mohan, D., & Varghese, M. (2008). Farm hand tools injuries: A case study from northern India. *Safety Science, 46*, 54–65. https://doi.org/10.1016/j.ssci.2007.03.003

Kurukulasuriya, P., & Ajwad, M. I. (2007). Application of the Ricardian technique to estimate the impact of climate change on smallholder farming in Sri Lanka. *Climatic Change, 81*, 39–59.

Kwon, O., Jung, K., You, H., & Kim, H. E. (2009). Determination of key dimensions for a glove sizing system by analyzing the relationships between hand dimensions. *Applied Ergonomics, 40*(4), 762–766.

Mariara, J. K., & Karanja, F. K. (2007). The economic impact of climate change on Kenyan crop agriculture: A Ricardian approach. *Global and Planetary Change, 57*, 319–330.

Mendelsohn, R., Nordhaus, W. D., & Shaw, D. (1994). The impact of global warming on agriculture: A Ricardian analysis. *The American Economic Review, 84*, 753–771.

Mishra, D., & Satapathy, S. (2018a). Hand tool injuries of agricultural farmers of South Odisha in India. *Materials Today: Proceedings, 5*, 17648–17653.

Mishra, D., & Satapathy, S. (2018b). Musculoskeletal & Risk Assessment during threshing of Rice grains. *MATEC Web of Conferences, 172*, 05005. https://doi.org/10.1051/matecconf/201817205005

Mittal, V. K., Bhatia, B. S., & Ahuja, S. S. (1996). A study of the magnitude, causes, and profile of victims of injuries with selected farm Machines in Punjab. Final report of ICAR adhoc research project. In *Department of Farm Machinery and Power Engineering*. Punjab Agricultural University.

Nelson, J. B., & Mital, A. (1995). An ergonomic evaluation of dexterity and tactility with increase in examination/surgical glove thickness. *Ergonomics, 38*(4), 723–733.

Nhemachena, C., & Hassan, R. (2007). *Micro-level analysis of farmers adaption to climate change in southern Africa*. International Food Policy Research Institute.

Rains, G. C., & Thomas, D. L. (2009). *Precision farming: An introduction*. University of Georgia.

Rautiainen, R. H., & Reynolds, S. J. (2002). Mortality and morbidity in agriculture in the United States. *Journal of Agricultural Safety and Health, 8*(3), 259–276. https://doi.org/10.13031/2013.9054

Reid, S., Smit, B., Caldwell, W., & Belliveau, S. (2007). Vulnerability and adaptation to climate risks in Ontario agriculture. *Mitigation and Adaptation Strategies for Global Change, 12*, 609–637.

Sørensen, C. G., Pesonen, L., Bochtis, D. D., Vougioukas, S. G., & Suomi, P. (2011). Functional requirements for a future farm management information system. *Computers and Electronics in Agriculture, 76*(2), 266–276. https://doi.org/10.1016/j.compag.2011.02.005

Tiwari, P. S., Gite, L. P., Dubey, A. K., & Kot, L. S. (2002). Agricultural Injuries in Central India: Nature, Magnitude, and Economic Impact. *Journal of Agricultural Safety and Health, 8*(1), 95–111.

Tremblay-Lutter, J. F., & Weihrer, S. J. (1996). Functional fit evaluation to determine optimal ease requirements in chemical protective gloves. *ASTM Special Technical Publication, 1237*, 367–383.

Turner, L. W., Udal, M. C., Larson, B. T., & Shearer, S. A. (2000). Monitoring cattle behavior and pasture use with GPS and GIS. *The Canadian Journal of Animal Science, 80*(3), 405–413.

Villa-Henriksen, A., Edwards, G. T. C., Pesonen, L. A., Green, O., & Aage Grøn Sørensen, C. (2020). Internet of things in arable farming: Implementation, applications, challenges and potential. *Biosystems Engineering, 191*, 60–84. https://doi.org/10.1016/j.biosystemseng.2019.12.013

Xiang, H., Wang, Z., Stallones, L., Keefe, T. J., Huang, X., & Fu, X. (2000). *American Journal of Public Health, 90*, 1269–1276.

Chapter 6
Food Safety and IoT-Based Solution

6.1 Introduction

The increasing agricultural production development has brought its significance in the global agri-sectors. So far agribusiness is concerned; it includes collective business-activities covering the agricultural inputs' supply, production as well as transformation of agri-products and their distribution to the end-users. It has been one of the key generators of employment in addition to income throughout the world. In addition to inadequate education, training, and unsuccessful development in local technologies, the confluence of a variety of local and global drivers has also led to progressing and unpredictable transforms in the functions and structures of worldwide agri-food markets. Global agricultural sectors in some rural areas are facing new challenges to find a market for the effective sale of their products. These problems of agricultural marketing are a serious problem worldwide. In addition, the challenges and opportunities presented by the trade regime liberalized from global markets are also difficult to find. Therefore, the transition to sustainable food and agriculture systems has become essential to effectively manage global agricultural and food markets, to support expected population growth, and to ensure global access to an adequate agricultural and food market in addition to safe and nutritious food.

This paper deals with finding the worldwide challenges associated with the food and agricultural systems, so that suitable intervention strategies can be undertaken at different levels of service and policy formulation in addition to competent governments. Moreover, in order of generating more profits, some businessmen do frauds by providing or supplying the old food that may decay one's life and without any tracing device, the stuffs or foods remain undetected as old and unhealthy or fresh and healthy one. So an IoT-based model is designed for providing "temperature, humidity and pH values" of the stored crops in any closed container for easier detection of the freshness of stored foods.

S. Satapathy et al., *Innovation in Agriculture with IoT and AI*,
Innovation, Technology, and Knowledge Management,
https://doi.org/10.1007/978-3-030-88828-2_6

6.2 Literature Review

Although different initiations were taken by many developing countries for the adoption of structural-adjustment policies in addition to widen poverty-reduction strategies for enhancing of economic-growth through market-opportunities, but it has resulted in semi-continuation agriculture owing to the engagement of huge number of small-scale farmers in these countries (Dorward & Kydd, 2004). The agricultural sub-sectors have been characterized in leading to higher transaction-costs along with pervasive market-imperfections because of weaker institutions and poorer infrastructures (Dorward et al., 2005; Kydd & Dorward, 2004). In addition, the positive-effects of liberalization were muted for the state marketing-agencies in agricultural market-sectors by partially implementing market-reforms, policy-reversals, and re-emergences (Jayne et al., 2002). A broader knowledge-system operates with agriculture extension that includes research in addition to agricultural-education. The World Bank and FAO refer this well-built-system as "Agricultural-Knowledge and Information-Systems for Rural-Development (AKIS/RD)", while these are referred as "Agricultural-Knowledge-System (AKS)" by the 'OECD countries' (FAO, 2001). Olesen and Bindi (2002) have concluded of the existing yield-gap for the former 'Soviet-countries in Europe' to be 4.5 times more than the probable growth in production from climatic-changes by 2050. For agribusiness orientations, the farming groups require considerable adherence to business-principles in addition to the overlapping-demands for societal inclusions (Chirwa et al., 2005).

Over two-thirds of all food retailing have been reported to be accounted in Western European as well as North American countries by the super-markets and hyper-markets Chen et al. (2005), which can be seen in other areas of the developing world such as South Asia and Latin America. However, this retail-revolution has brought new opportunities as well as barriers for the agri-food systems. Dolan and Humphrey (2000) have studied about the fresh-food exporting from Africa, and noted of UK super-markets having established-control over the fresh-vegetables trade with a decisive-role to determine the production as well as processing structures of exported fresh-vegetables from Africa. In a study by ICRIER commissioned by 'Ministry of Commerce and Industry', India, for the analysis of the impacts of organised-retailing on unorganised-retail, farmers and mediators, it was found of coexisting of both and the organised-retail would contribute by setting-up modern supply-chains with growth in productivity for both industry and agriculture (Joseph et al., 2008). Owing to the inabilities of poorer farmers in developing countries in adoption of 'technological innovations' as a result of limited-resources with smaller access to information and markets, they are likely be excluded from trade (Ruben et al., 2006). The successful fresh-food firm understands the consumer-value drivers for the targeted market with execution of suitable strategies for creating and delivering of values to end users. The suppliers with knowledge of products, including the end-use by consumers and sensitive to the market-demands, will be able in responding more quickly to varying consumer-preferences and in retaining of marketing advantages (Tropp et al., 2008).

In order to understand the product-flows, exchange-levels, and the affecting forces for the supply-chain operation, detailed study on the existing trade-system as well as trade-environment become essential, which can help in identifying and delineating the functions of the potential supply-chain partners of an agri supply-chain. It is necessary to establish cooperation among supply-chain partners in case of food-safety problems for improvement in quality verification and reduction in costs. In a study of supply-chain regarding 'malting barley to beer agri-food', Leat and Reoldo-Giha (2008) have identified five-factors affecting the supply-chain relationships, such as communications, compatibilities of aims, commercial-benefits, contractual-relationships on the basis of professional-regard and personal-relationships, and higher trust-levels and willingness for resolving any problems, respectively.

The activities involved in food-processing are usually time-consuming and labour-intensive resulting in lower-yields and poorer economic-returns. Hence, identifying of these challenges becomes essential to assist in overcoming them for future capacity-building, value-addition, and establishment of successful in addition to sustainable businesses (Mercer, 2011). The diffusion of enhanced agricultural-technology and management-practices in different parts of the world can be traced-back thousands of years, including Egypt, Mesopotamia, China, and in America also. The origin of public/government funded extension-systems can be traced-back to the United Kingdom as well as Ireland, and during the scarcity period of potatoes in Ireland in between 1845 to 1851, the Irish potato-farmers were helped by the agricultural extension-advisors to diversify into different food-crops (Swanson & Rajalahti, 2010). The extension-service performs the role of developing practical research-knowledge applications, and providing of instructions and improved technologies or practices in agriculture (FAO, 1997). Six major-streams of both rural and urban areas have been reported to be covered by the U.S. extension, such as youth-development; agriculture; leadership-development; natural-resources; consumer and family sciences; and economic and community development; respectively (USDA, 2011).

There has been a growing-concern about environmental and social sustainability of food-industries (Li et al., 2014). Our global food-production system may not meet the increasing resources intensive-consumption patterns, and if the existing trends of population in addition to consumption continue, then "by 2030, the civilization will call for the equivalence of Two-Earths for supporting" (Moomaw et al., 2012). Agricultural marketing helps in the reformation as well as creation of marketing-infrastructures. Therefore, it becomes essential in expanding all the services that will develop agricultural-marketing related to improvement in marketing-system, strengthening the marketing-infrastructures, investment requirements, possible fund-sources, improving marketing information-system by the use of "Internet Communication Technology (ICT)", human-resource development in agricultural marketing, and exports related promotional measures (Kusugal & Nagaraja, 2013). With regard to different empirical-studies in "Asia, Africa, and Latin America", the producer organizations' role has been documented showing the smallholders' capabilities in accessing to higher value-markets (Bernard et al.,

2008; Bernard & Spielman, 2009; Fischer & Qaim, 2012; Hellin et al., 2009; Narrod et al., 2009; Kaganzi et al., 2009; Shiferaw et al., 2011; Trebbin, 2014; Wollni & Zeller, 2007; Wollni et al., 2010).

The adoption intensities of farmers in Tanzania for improved varieties of maize was found to be influenced with farmer-groups' membership (Amare et al., 2012). Studies have recommended of sustainable agriculture practices through tillage-conservation, diversification in cropping-systems and investments in conserving of water as well as soil (Kassie et al., 2013; Teklewold et al., 2013). The conceptual-framework of value-chains' in agriculture includes value-addition activities including "production, processing, marketing and consumption". According to Food and Agricultural Organisation, the terms supply-chain and value-chain are often interchanged (FAO, 2005). The value-chain can be a network or vertical linking between various independent business-organisations comprising of "processing, packaging, storing, transporting and distributing". In South Asian countries, the agricultural value-chains are often disjointed with missing linkages of markets and farms (Dunn, 2014). The rural areas in the world account for 75% poor-individuals according to World Bank (2015), and they mainly depend on agriculture as key income source. Peoples below poverty-levels of these rural areas also remain higher and stubborn in South Asia. The substantial growth of agriculture can be an effective-tool for the reduction in poverty of rural areas similar to other sectors' growth in economy-levels. Shiferaw et al. (2016) have suggested of democratic-governance, consistent and most favourable group-size, transparency, and market-orientation to improve the agricultural producer-organizations' performances. Markets can be regarded as collective mechanism allowing give and take philosophy confined not only to products and distribution, but also provide the necessary values (Callon & Muniesa, 2005), which are shaped by cultural-customs, public-regulations, civic-norms and private -contracts (Callon, 1998).

Moreover, the agri-food markets play a vital-role in the aggregation of demands and supplies throughout the entire food-system starting from input-supply to farm-production, collections, processing as well as packaging, transporting and end-consumption of retail food-products (Barrett & Mutambatsere, 2008; Myers et al., 2010), thus it makes the final possibility for crops to be consumed (Saccomandi & van der Ploeg, 1998; Sodano et al., 2010), and initiates in economic-development (Barrett & Mutambatsere, 2008). There has been a rapid global transformation in the agri-food markets with the global-drivers creating of certain meta-trends significantly affecting the agri-food sectors throughout the world (Narayanan & Gulati, 2002; Reardon, 2000). Nowadays, the agri-food markets and systems have been reported to be under a wide range of pressures (FAO, 2014). In fact, they are at the interconnection-centre of worldwide social, environmental as well as economic problems having critical-deal with various sustainability challenges such as climate-changes, population-growths, food insecurities and malnutrition, resources-scarcities, ecosystem-degradations, and biodiversity-losses (FAO, 2014; Foresight, 2011; Garnett, 2014; Gladek et al., 2016; Godfray et al., 2010; Lang, 2009; Searchinger et al., 2013; Vermeulen et al., 2012; Yakovleva, 2007). Different authors have argued of globalization having a positive-impact on food-security by risk reductions of global supply-shock (Runge et al., 2003), food-prices (FAO, 2004), and their volatilities (World Bank, 2007). Over the last decades, the food safety and quality standards have meaningfully transformed the global agri-food value-chains (Henson & Reardon, 2005; FAO, 2015).

The globalisation of agricultural value-chains creates a principal challenge in serving local, national in addition to international markets. According to the "United Nations Conference on Trade and Development" majority of companies were found of getting reluctant in cooperating with the local-farmers as a consequence of structural-shortcomings, such as poorer reliability and lack of quality-products (UNCTAD, 2006). Fruits' processing accounted for 70% in Brazil; 50% in Indonesia; 23% in China; and only 4% in India (Shivakumar, 2016). Sundar (2016) has explored more than fifty agri-business related opportunities in India with an insight to the constraints associated with such business. The advancement of global-economy has been considerably through digitalization with the digital-market as the most established and prominent platform in developed-markets having 75 to 80% share in total-marketing than developing-economies (Ismagilova et al., 2017). Digital-agriculture could be a prime requirement to serve the world by 2050 with expected population of 9.7 billion that can be possible with precision-farming adoption through digitalisation (G Global Innovation Index, 2017). At present the digital-platforms are mostly used in four agricultural areas such as "Website-development, Social-media marketing, Mobile-app development, and Content-marketing", respectively (Hussmann, 2018). There has been a higher pressure on businesses as well as governments for paying more attention on the resource and environmental consequences of the ever-growing productions, distributions and consumptions of 'agro-based products'. The active participation of producers like small-scale farmers in sourcing-networks in addition to the institutional-initiatives can help in the sustainable development by meeting the stringent food-safety as well as quality-regulations (Naik & Suresh, 2018). Borsellino et al. (2020) have suggested for further extension of research for the improvement of basic-knowledge and in the identification of opportunities in order to design meaningful-actions, which can shape the agri-food markets towards sustainability.

The four dimensions or pillars of food-security include food availability, accessibility, consumption and stability over-time (FAO, 2006), and meeting of global food-security requirements remain as a challenge with the enhanced demands for foods with the population-growth. There is a higher requirement for addressing the ever increasing stresses on food-security and quality food-demands to ensure sustainably by minimizing the environment-impacts and maximization of social-opportunities (Hobbs, 2007). The "Food and Agriculture Organization" of the United Nations report identified the provision-requirements for sufficient and reasonable food-supplies through sustainable agriculture services with the intention of meeting the rising demands of the world-population (FAO, 2017).

Regardless of the decreased population growth-rates at the international-level, South Asia and Africa will still witness a significant amplification in the coming-years that will cause bigger competition for already limited land as well as water-resources (Lammers et al., 2000). In addition, the competitions for natural-resources inputs for foods and energy utilizations (Harvey & Pilgrim, 2011), and strategies for national bio-economy (Bracco et al., 2018) will also likely to increase. However, apart from the pressures on land-resources, the scarcity of water due to industrial, agricultural and urban requirements account for over 40% of worldwide rural population (FAO, 2018a). One of the premises for intensification of sustainability, the productivity of the agricultural sectors need enhancement for meeting the demand growth-rates (Garnett et al., 2013); and based on past study, the productivity in 2050 need to be increased to almost 50% more than the year 2012 (World Bank Group, 2016). More agriculture output would be required under an increasing food-wastes scenario, income-inequality, dissimilarity in food-distributions, and worsen climate-changes (FAO, 2018b). The climatic-changes not only deteriorate the livelihood, but also pose direct-risks on food-security with reduced employment-opportunities as well as poverty-levels, and limit the access to social-protections (FAO, 2016).

In view of the above literature, the objectives of this study were set as follows:

- To identify the associated challenges with the worldwide food and agricultural systems;
- To prioritize the worldwide associated challenges based on their preferences of impacts' on agriculture.
- To propose and develop an IoT-based model for dealing with food-safety.

Moreover, for solving any "Multi-Criteria Decision-Making (MCDM)" problem, different studies have recommended the efficient-use of "Step-Wise Assessment and Ratio-Analysis (SWARA)" method (Alimardani et al., 2013; Mishra & Satapathy, 2020a; Mishra & Satapathy, 2020b; Yazdani et al., 2016; Zolfani & Bahrami, 2014; Zolfani et al., 2018; Zolfani & Chatterjee, 2019). Thus, this method was chosen in this study for analysing and ranking of the associated challenges with the worldwide food and agricultural systems.

6.3 Research Methodology

This study included an extensive literature-review followed by consultation with agricultural, environmental and academic experts in the identification and subsequent prioritization of the associated challenges with the worldwide food and agricultural systems. The SWARA method was utilized for the ranking of the challenges based on their preferences for prioritization. Finally, an IoT-based device was proposed as a food safety measure.

The participated experts' in the process of decision-making with their domain expertise-areas, gender, educational-qualifications, and average-years of experiences for this study were illustrated in Table 6.1.

Table 6.1 Participated experts' characteristics

Domain expertise-areas	Gender		Educational-qualifications		Average-years of experiences
	Male	Female	Master-degree	Doctorate	
Agricultural	5	1	0	6	More than 20
Environmental	3	0	0	3	More than 20
Academic	4	2	1	5	More than 25

Table 6.2 The associated challenges (AC) for the worldwide food and agricultural systems

Sl. No.	Associated major-challenges	Associated sub-challenges
1	Challenges associated with food availability and stability (AC_1)	Prevention of trans-boundary and emerging food as well as agricultural systems threats (AC_{1-1})
		Meeting of increasing demands through sustainable improvement in agricultural productivity (AC_{1-2})
		Addressing of climatic-changes and intensification of natural-hazards (AC_{1-3})
		Ensuring of sustainable natural resource-base (AC_{1-4})
2	Challenges associated with food-access and their utilizations (AC_2)	Ending of hunger and other related malnutrition (AC_{2-1})
		Elimination of acute poverty and reduction of inequalities (AC_{2-2})
		Building up of resilience for the prolonged conflicts, crises, and disasters (AC_{2-3})
		Improvement in rural areas' income-opportunities with addressing of the root-causes for migration (AC_{2-4})
3	General challenges (AC_3)	Addressing of the needs for effective and coherent governance at national and international levels (AC_{3-1})
		Making the food-systems more efficient, resilient, and inclusive (AC_{3-2})

6.3.1 *Identification of the Associated Challenges for the Worldwide Food and Agricultural Systems*

The potential and possible associated challenges for the worldwide food and agricultural systems were grouped into three major-groups (FAO, 2017), such as: (a) challenges associated with food availability and stability; (b) challenges associated with food-access and their utilizations; and (c) general challenges; illustrated in Table 6.2 with the associated sub-challenges, respectively.

6.3.2 Steps-Involved in 'SWARA Method'

The ranking of most-important as well as least-important criteria (the associated challenges in this study) at higher and lower levels by the use of SWARA method was based on the following steps (Keršuliene et al., 2010; Mishra & Satapathy, 2020b; Stanujkic et al., 2015):

Step-A: Sorting of criteria on the basis of importance

In this step the ranking of criteria was done by the experts based on the relative-importance of each criterion. The placing of most-important criteria at the first-place was done initially, while was followed by the placing of least-important criteria at the last-place.

Step-B: Determining the relative-importance of average-values (s_j)

Starting from the second-ranked criterion, the relative-importance of average-values (s_j) were determined based on the importance of criterion (c_j) than the criterion (c_{j+1}).

Step-C: Calculating the coefficients (k_j) as follows:

$$k_j = \begin{cases} 1, j = 1 \\ s_j + 1, j = 1 \end{cases}$$

Step-D: Determining the recalculated-weights (q_j) as follows:

$$q_j = \begin{cases} 1, j = 1 \\ \dfrac{q_{j-1}}{k_j}, j > 1 \end{cases}$$

Step-E: Determining the relative-weights (w_j) of the evaluation-criteria as follows:

$$w_j = \frac{q_j}{\sum_{k=1}^{n} q_k}$$

Where, n = Number of criteria.

6.4 Results

The associated major-challenges in addition to the corresponding sub-challenges for the worldwide food and agricultural systems, as illustrated in Table 6.2 were considered as criteria and sub-criteria for subsequent ranking by SWARA method. The identified three-criteria included "Challenges associated with food availability and stability (AC_1); Challenges associated with food-access and their utilizations (AC_2); and General challenges (AC_3)", respectively. While, the four sub-criteria under the criteria of "challenges associated with food availability and stability" included "Prevention of trans-boundary and emerging food as well as agricultural systems threats (AC_{1-1}); Meeting of increasing demands through sustainable improvement in agricultural productivity (AC_{1-2}); Addressing of climatic-changes and intensification of natural-hazards (AC_{1-3}); and Ensuring of sustainable natural resource-base (AC_{1-4})". The four identified sub-criteria under the criteria of "challenges associated with food-access and their utilizations" included "Ending of hunger and other related malnutrition (AC_{2-1}); Elimination of acute poverty and reduction of inequalities (AC_{2-2}); Building up of resilience for the prolonged conflicts, crises, and disasters (AC_{2-3}); and Improvement in rural areas' income-opportunities with addressing of the root-causes for migration (AC_{2-4})". Correspondingly, the two sub-criteria under the criteria of "general challenges" included "Addressing of the needs for effective and coherent governance at national and international levels (AC_{3-1}); and Making the food-systems more efficient, resilient, and inclusive (AC_{3-2})"; respectively.

The corresponding criteria weights' and sub-criteria weights' were calculated based on the SWARA method, and subsequently their ranking was done. The weights of criteria as well as sub-criteria were shown in Table 6.3, Table 6.4, Table 6.5, and Table 6.6, respectively. The relative-importance of average-values (s_j)

Table 6.3 Final-weights' of criteria AC_1, AC_2 and AC_3

Criteria	Relative-importance of average-values (s_j)	Coefficients ($k_j = s_j + 1$)	Recalculated-weights $\left(w_j = \frac{x_{j-1}}{k_j}\right)$	Weight $\left(q_j = \frac{w_j}{\sum w_j}\right)$
AC_2		1	1	0.400
AC_1	0.23	1.23	0.813	0.325
AC_3	0.19	1.19	0.683	0.273

Table 6.4 Final-weights' of sub-criteria for AC_1

Sub-Criteria	Relative-importance of average-values	Coefficients	Recalculated-weights	Relative-importance of average-values	Final-weights
AC_{1-1}		1	1	0.312	0.101
AC_{1-3}	0.18	1.18	0.847	0.264	0.085
AC_{1-2}	0.17	1.17	0.724	0.225	0.073
AC_{1-4}	0.14	1.14	0.635	0.198	0.064

Table 6.5 Final-weights' of sub-criteria for AC_2

Sub-Criteria	Relative-importance of average-values	Coefficients	Recalculated-weights	Relative-importance of average-values	Final-weights
AC_{2-3}		1	1	0.307	0.122
AC_{2-1}	0.17	1.17	0.854	0.262	0.104
AC_{2-2}	0.14	1.14	0.749	0.230	0.092
AC_{2-4}	0.15	1.15	0.651	0.200	0.080

Table 6.6 Final-weights' of sub-criteria for AC_3

Sub-Criteria	Relative-importance of average-values	Coefficients	Recalculated-weights	Relative-importance of average-values	Final-weights
AC_{3-1}		1	1	0.537	0.146
AC_{3-2}	0.16	1.16	0.862	0.463	0.126

for both criteria as well as sub-criteria was obtained by the help of experts. The weights of criteria were utilized to obtain the final-weights' of concerned sub-criteria. For the calculations, the assessment-scale was based on 5% steps, such as the experts' indicated a comparative and value-difference on the basis of 5% increments including 5%; 10%; 15%; and so on. However, the final-numbers in this part were considered from the arithmetic-average of experts' opinions.

Subsequently, weights of all the concerned criteria and sub-criteria together with their relative-ranking with respect to the values of their final-weights' were summarized in Table 6.7.

6.5 Discussion

In this study on the basis of prioritization of criteria i.e. major-challenges for the worldwide food and agricultural systems, it was found from the ranking that "Challenges associated with food-access and their utilizations" ranked first followed by "Challenges associated with food availability and stability"; and "General challenges"; respectively. Similarly, based on the prioritization of sub-criteria as sub-challenges for all the concerned criteria, the corresponding ranking in the descending order included "Addressing of the needs for effective and coherent governance at national and international levels; Making the food-systems more efficient, resilient, and inclusive; Building up of resilience for the prolonged conflicts, crises, and disasters; Ending of hunger and other related malnutrition; Prevention of trans-boundary and emerging food as well as agricultural systems threats; Elimination of acute poverty and reduction of inequalities; Addressing of climatic-changes and intensification of natural-hazards; Improvement in rural areas' income-opportunities with addressing of the root-causes for migration; Meeting of increasing

Table 6.7 Final-summary of the weights of criteria and sub-criteria

Criteria and Sub-criteria	Final-weights	Final-weight's based criteria rank	Final-weight's based sub-criteria rank
Challenges associated with food availability and stability (AC_1)	0.325	**2**	--
Prevention of trans-boundary and emerging food as well as agricultural systems threats (AC_{1-1})	0.101	--	**5**
Meeting of increasing demands through sustainable improvement in agricultural productivity (AC_{1-2})	0.073	--	**9**
Addressing of climatic-changes and intensification of natural-hazards (AC_{1-3})	0.085	--	**7**
Ensuring of sustainable natural resource-base (AC_{1-4})	0.064	--	**10**
Challenges associated with food-access and their utilizations (AC_2)	0.400	**1**	--
Ending of hunger and other related malnutrition (AC_{2-1})	0.104	--	**4**
Elimination of acute poverty and reduction of inequalities (AC_{2-2})	0.092	--	**6**
Building up of resilience for the prolonged conflicts, crises, and disasters (AC_{2-3})	0.122	--	**3**
Improvement in rural areas' income-opportunities with addressing of the root-causes for migration (AC_{2-4})	0.080	--	**8**
General challenges (AC_3)	0.273	**3**	--
Addressing of the needs for effective and coherent governance at national and international levels (AC_{3-1})	0.146	--	**1**
Making the food-systems more efficient, resilient, and inclusive (AC_{3-2})	0.126	--	**2**

[a]'Final-weights' of sub-criteria, $AC_{2-1} = 0.400 \times 0.262 = 0.104$; and $AC_{3-1} = 0.273 \times 0.537 = 0.146$

demands through sustainable improvement in agricultural productivity; and Ensuring of sustainable natural resource-base; respectively.

However, the increasing flexibility of the food-systems would require adequate-solutions beyond technological improvements with a broader-focus to enhance the productivity in agricultural sectors, along with sufficient consideration of the trade-offs with the environmental changes (Ericksen et al., 2009). Worldwide, greater crop diversity was accounted by the small-scale farmers having below two-hectares of farm-land and producing 30 to 34% of the world's food-supply from a gross agricultural-area of 24% (Ricciardi et al., 2018). Therefore, the institutions, business-models, and policies that provide support and incentives for linkage of small-scale farmers to markets are much essential. Moreover, there occurs a considerable amount of globally produced food to be lost and/or wasted, which causes higher environmental-burden in addition to inefficient utilization of resources

(Foley et al., 2011). Thus, efficient food-systems would be the essential solution to such problems.

This study confirms with the reports by FAO (2018a) and FAO (2018b), which highlighted stronger linkage of climate-changes with other trends in the 'food and agriculture' sectors, such as the development requirement for efficient as well as flexible food-systems (like intensification of agriculture, avoidance of land-degradation, more efficient utilization of water, and so on), and reduction of inequalities through effective and coherent governance-systems.

The findings of SWARA ranking resulted to provide more attention on the "challenges associated with food-access and their utilizations" for the worldwide food and agricultural systems. Moreover, different researches have been done on food quality and safety (Annese & De Venuto, 2015; Brody et al., 2008; Pang et al., 2010). However, it was a long process in determining the quality of food as good or not. By considering as well as testing of a variety of factors, they concluded with the food's quality to be healthier to take or not. Thus, there is a higher need of some device to make the food quality detection process easier and faster.

6.5.1 Need of Food Safety Measures

In today's business supply chain, it becomes a major concern to grow or run the business successfully. Food is the main product to live, so supply chain management in food has been of significance importance nowadays. However, to make profits, some businessmen do frauds by providing or supplying the old food that decays one's life. Therefore, tracing of food items is of prime importance to avoid the final consumption of such unhealthy foods by the individuals.

Food safety and hygiene is a major concern in order to prevent the food wastage. The quality of the food needs to be monitored and it must be prevented from rotting and decaying by the atmospheric factors like temperature, humidity and others. Therefore, it is useful to deploy quality monitoring devices at food stores. These quality monitoring devices keep a watch on the environmental factors that cause decay of the food. Further, the environmental factors can be controlled like by refrigeration, vacuum storage, and so on.

6.5.2 Design Concept

An IoT-based model is designed, which will show the real time temperature as well as the humidity of a closed container with crops. The block-diagram of the proposed device and the corresponding operational flow chart is shown in Fig. 6.1 and Fig. 6.2, respectively. The users of this device need to have prior knowledge regarding the limits of temperature and humidity levels of food items to remain fresh. This device can provide the temperature, humidity and pH value's readings (with additional attachments) that will help in the detection of whether the corps are old or

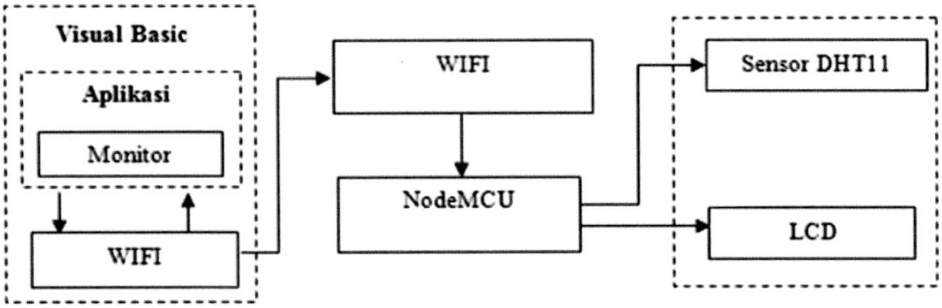

Fig. 6.1 Block-diagram of the proposed device

Fig. 6.2 Operational flow chart of proposed device

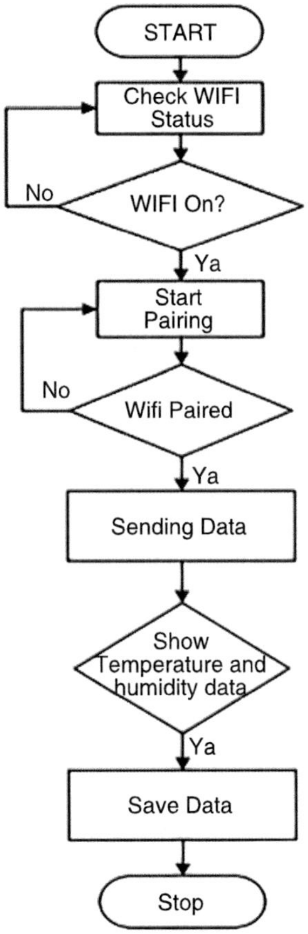

fresh. As a main component, DHT 11 is used in this design. Furthermore, this design can also help to justify the fruits and bakery things (like: Cake, Biscuits, etc.). It has also great future implements, like by connecting a gas detector, it is possible to determine of the amount and type of gas formation after the end of freshness of the food. This proposed device has a greater opportunity to grow, as it can help common people from frauds.

6.5.3 *Working Principle and Future Scope of the Proposed Device*

The wireless sensor unit is used to monitor the critical environmental parameters like temperature, humidity, moisture etc. (Fig. 6.3). The use of DHT-11 sensor helps to sense the humidity and temperature, and gives an analog value to node micro-controller unit (MCU). As input is given to MCU through laptop or mobile-phones, it changes the analog values into digital values. A LCD screen is attached with the MCU to show the digital values.

A 24bit ADC 'load cell HX711' is used to amplify and for digital conversion of the Load cell output. Then this amplified value is fed to the Arduino. The Arduino calculates the output of HX711, and converts that into the weight values in grams and shows it on LCD. A push-button is used for calibrating the system. The food quality test by the use of proposed device is shown in Fig. 6.4.

With the connection of more sensors like gas indicator and pH sensor, the foods quality can be measured more efficiently. Further, as less number of load cells are used to measure small weight of food, it may be enhanced with more load to measure the quality of larger quantity of foods at a time.

6.6 Conclusion

In this study, by the use of SWARA method and subsequent prioritization of criteria i.e. major-challenges for the worldwide food and agricultural systems, it was found that "Challenges associated with food-access and their utilizations" ranked first followed by "Challenges associated with food availability and stability"; and "General challenges"; respectively. Similarly, based on the prioritization of sub-criteria as

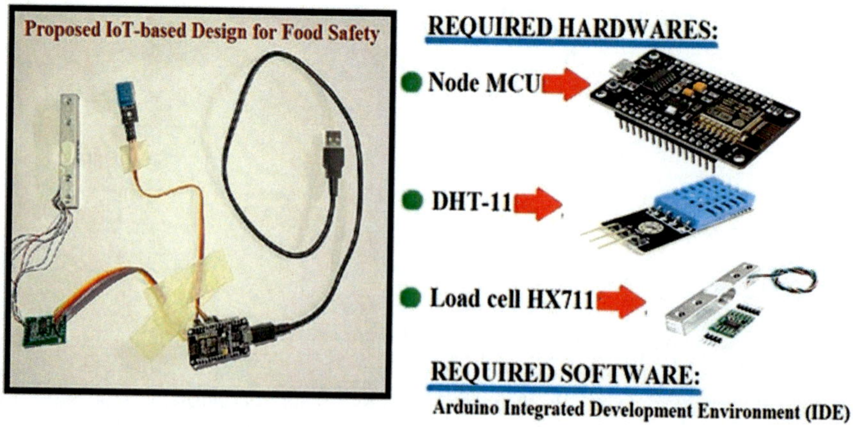

Fig. 6.3 Food safety measuring device

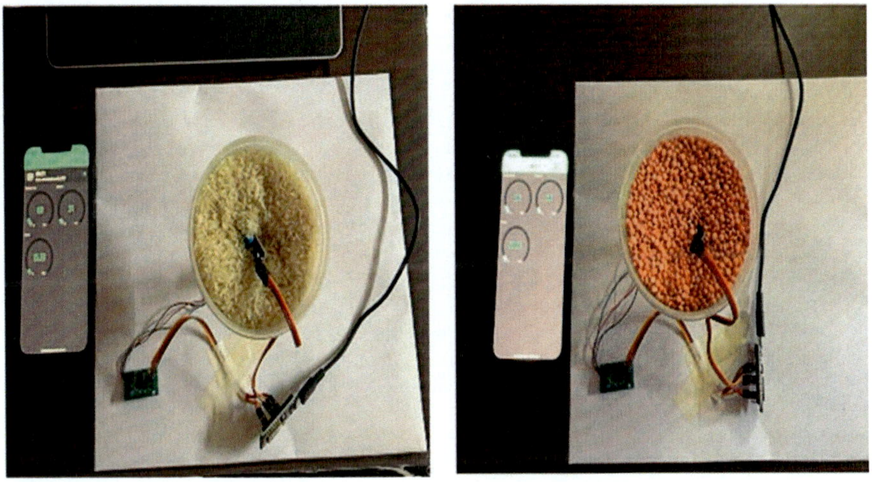

Fig. 6.4 Food quality test

sub-challenges for all the concerned criteria, the corresponding ranking in the descending order included "Addressing of the needs for effective and coherent governance at national and international levels; Making the food-systems more efficient, resilient, and inclusive; Building up of resilience for the prolonged conflicts, crises, and disasters; Ending of hunger and other related malnutrition; Prevention of trans-boundary and emerging food as well as agricultural systems threats; Elimination of acute poverty and reduction of inequalities; Addressing of climatic-changes and intensification of natural-hazards; Improvement in rural areas' income-opportunities with addressing of the root-causes for migration; Meeting of increasing demands through sustainable improvement in agricultural productivity; and Ensuring of sustainable natural resource-base; respectively.

Moreover, the proposed design for the food safety measures can be very helpful for the retailers buying smaller quantities of food. This device can measure the temperature, humidity and weight at a time, and also capable of showing the output in a single display. So it will be more convenient to the buyers in the assessment of the quality and quantity of food at a glance. Further, with the addition of some more usable devices, it will be great gadget for food supply chain.

References

Alimardani, M., Zolfani, S. H., Aghdaie, M. H., & Tamošaitienė, J. (2013). A novel hybrid SWARA and VIKOR methodology for supplier selection in an agile environment. *Technological and Economic Development of Economy, 19*(3), 533–548.

Amare, M., Asfaw, S., & Shiferaw, B. (2012). Welfare impacts of maize–pigeonpea intensification in Tanzania. *Agricultural Economics*, 1–17.

Annese, V., & De Venuto, D. (2015). On-line shelf-life prediction in perishable goods chain through the integration of WSN technology with a 1st order kinetic model. In *2015 IEEE 15th International Conference on Environment and Electrical Engineering, EEEIC 2015 – Conference Proceedings*. https://doi.org/10.1109/EEEIC.2015.7165232

Barrett, C. B., & Mutambatsere, E. (2008). *Agricultural markets in developing countries* (pp. 91–95). In The New Palgrave Dictionary of Economics; Nature Publishing Group.

Bernard, T., & Spielman, D. (2009). Reaching rural poor through rural producer organizations? A study of the agricultural marketing cooperatives in Ethiopia. *Food Policy, 34*, 60–69.

Bernard, T., Taffesse, S. A., & Gabre-Madhin, E. (2008). Impact of cooperatives on smallholders' commercialization behaviour: Evidence from Ethiopia. *Agricultural Economics, 39*, 147–161.

Borsellino, V., Schimmenti, E., & Bilali, H. E. (2020). Agri-food markets towards sustainable patterns. *Sustainability, 12*, 2193. https://doi.org/10.3390/su12062193

Bracco, S., Calicioglu, O., Juan, M. G. S., & Flammini, A. (2018). Assessing the contribution of bioeconomy to the total economy: A review of national frameworks. *Sustainability, 10*, 1698.

Brody, A., Bugusu, E., Han, J., Sand, C., & Mchugh, T. (2008). Innovative food packaging solutions. *Journal of Food Science, 73*.

Callon, M. (1998). An essay on framing and overflowing: Economic externalities revisited by sociology. In M. Callon (Ed.), *The Laws of the markets* (pp. 244–269). Blackwell Publishing Ltd.

Callon, M., & Muniesa, F. (2005). Peripheral vision economic markets as calculative collective devices. *Organization Studies, 26*, 1229–1250.

Chen, K. Z., Shepherd, A. W., & da Silva, C. (2005). *Changes in food retailing in Asia—Implications of supermarket procurement practices for farmers and traditional marketing systems*. FAO.

Chirwa, E., Dorward, A., Kachule, R., Kumwenda, I., Kydd, J., Poole, N., Poulton, C., & Stockbridge, M. (2005). *Walking tightropes: Supporting farmer organizations for market access* (p. 99). Natural Resource Perspectives.

Dolan, C., & Humphrey, J. (2000). Governance and trade in fresh vegetables: The impact of UK supermarkets on the African horticulture industry. *The Journal of Development Studies, 37*(2), 147–176.

Dorward, A., & Kydd, J. (2004). The Malawi 2002 food crisis: The rural development challenge. *Journal of Modern African Studies, 423*, 343–361.

Dorward, A., Kydd, J., Morrison, J., & Poulton, C. (2005). Institutions, markets and economic development: Linking development policy to theory and praxis. *Development and Change, 361*, 1–25.

Dunn, E. (2014). *Smallholders and inclusive growth in agricultural value chains*. Field Report, 18, USAID. Accessed on November 15, 2020 from http://www.fhi360.org/sites/default/files/media/documents/FIELD%20Report%20No%2018%20Smallholders%20in%20Value%20Chains.pdf.

Ericksen, P. J., Ingram, J. S. I., & Liverman, D. M. (2009). Food security and global environmental change: Emerging challenges. *Environmental Science & Policy, 12*, 373–377.

FAO. (1997). *Improving agricultural extension. A reference manual*. Edited by Swanson, B., Bentz, R., and Sofranko, A. Rome, Italy. Available online at http://www.fao.org/docrep/W5830E/w5830e00.htm. Accessed 15 Nov 2020.

FAO. (2001). *Agricultural and Rural Extension Worldwide: Options for Institutional Reform in the Developing Countries, Rome, Italy*. Available online at ftp://ftp.fao.org/docrep/fao/004/y2709e/y2709e.pdf. Accessed 15 Nov 2020.

FAO. (2004). *The state of agricultural commodity markets 2004*. Rome, Italy.

FAO. (2005). *Addressing marketing and processing constraints that inhibit agrifood exports: A guide for policy analysts and planners*. FAO agricultural services bulletin, 60, Rome. Accessed on 15 Nov 2020 from http://www.fao.org/publications/card/en/c/08b594d3-ccb0-5463-a243-bd0cee835fee/

FAO. (2006). *Food security*. Policy Br. 2006.

FAO. (2014). *Building a common vision for sustainable food and agriculture—Principles and approaches*. Rome, Italy.

FAO. (2015). *The state of agricultural commodity markets 2015–16—Trade and food security: Achieving a better balance between National Priorities and the collective good*. Rome, Italy.
FAO. (2016). *Migration and Protracter crises*. Rome, Italy.
FAO. (2017). *The future of food and agriculture: Trends and challenges*. Rome, Italy. ISBN: 9789251095515.
FAO. (2018a). *The state of the World's land and water resources for food and agriculture: Managing Systems at Risk*. Rome, Italy.
FAO. (2018b). *The future of food and agriculture alternative pathways to 2050*. Rome, Italy.
Fischer, E., & Qaim, M. (2012). Linking smallholders to markets: Determinants and inputs of farmer collective action in Kenya. *World Development, 40*(6), 1255–1268.
Foley, J. A., Ramankutty, N., Brauman, K. A., Cassidy, E. S., Gerber, J. S., Johnston, M., Mueller, N. D., O'Connell, C., Ray, D. K., West, P. C., et al. (2011). Solutions for a cultivated planet. *Nature, 478*, 337–342.
Foresight (2011) The future of food and farming. Final project report; the government Office for Science: .
Garnett, T. (2014). Three perspectives on sustainable food security: Efficiency, demand restraint, food system transformation. What role for life cycle assessment? *Journal of Cleaner Production, 73*, 10–18.
Garnett, T., Appleby, M. C., Balmford, A., Bateman, I. J., Benton, T. G., Bloomer, P., Burlingame, B., Dawkins, M., Dolan, L., Fraser, D., et al. (2013). Sustainable intensification in agriculture: Premises and policies. *Science, 341*, 33–34.
Gladek, E., Fraser, M., Roemers, G., Sabag Munoz, O., Hirsch, P., & Kennedy, E. (2016). *The Global food system: An analysis*. Amsterdam, The Netherlands.
Global Innovation Index, G. (2017). *Innovation feeding the world*. Cornell University.
Godfray, H. C. J., Beddington, J. R., Crute, I. R., Haddad, L., Lawrence, D., Muir, J. F., Pretty, J., Robinson, S., Thomas, S. M., & Toulmin, C. (2010). Food security: The challenge of feeding 9 billion people. *Science, 327*, 812–818.
Harvey, M., & Pilgrim, S. (2011). The new competition for land: Food, energy, and climate change. *Food Policy, 36*, S40–S51.
Hellin, J., Lundy, M., & Meijer, M. (2009). Farmer organization, collective action and market access in Meso-America. *Food Policy, 34*, 16–22.
Henson, S., & Reardon, T. (2005). Private Agri-food standards: Implications for food policy and the Agri-food system. *Food Policy, 30*, 241–253.
Hobbs, P. R. (2007). Conservation agriculture: What is it and why is it important for future sustainable food production? *The Journal of Agricultural Science, 145*, 127–137.
Hussmann, S. (2018). *Automation in agriculture: Securing food supplies for future generations*.
Ismagilova, E., Dwivedi, Y. K., Slade, E., Williams, M. D. (2017) *Electronic Word of Mouth (eWOM) in the Marketing Context: A State of the Art*.
Jayne, T. S., Govereh, J., Mwavumo, A., Nyoro, J. K., & Chapoto, A. (2002). False promise or false premise? The experience of food and input market reform in Eastern and Southern Africa. *World Development, 3011*, 1967–1985.
Joseph, M., Soundararajan, N., Gupta, M., Sahu, S. (2008) *Impact of organized retailing on the unorganized sector*. Indian Council for Research on International Economic Relations (ICRIER).
Kaganzi, E., Ferris, S., Barham, J., Abenakyo, A., Sanginga, P., & Njuki, J. (2009). Sustaining linkages to high value markets through collective action in Uganda. *Food Policy, 34*, 23–30.
Kassie, M., Jaleta, M., Shiferaw, B., Mmbando, F., & Mekuria, M. (2013). Adoption of interrelated sustainable agricultural practices in smallholder systems: Evidence from rural Tanzania. *Technological Forecasting and Social Change, 80*, 525–540.
Keršuliene, V., Zavadskas, E. K., & Turskis, Z. (2010). Selection of rational dispute resolution method by applying new step-wise weight assessment ratio analysis (SWARA). *Journal of Business Economics and Management, 11*(2), 243–258.

Kusugal, P. S., & Nagaraja, S. (2013). Opportunities & challenges for agriculture marketing. *Indian Journal of Applied Research, 3*(8), 44–45.

Kydd, J., & Dorward, A. (2004). Implications of market and coordination failures for rural development in least developed countries. *Journal of International Development, 16*, 951–970.

Lammers, R. B., Salisbury, J., Green, P., & Vörösmarty, C. J. (2000). Global water resources: Vulnerability from climate change and population growth. *Science, 289*, 284–288.

Lang, T. (2009). *Food security and sustainability: The perfect fit*. London, UK.

Leat, P., Reoldo-Giha, C. (2008) *Enhancing the integration of agri-food supply chains: Theoretical issues and practical challenges in UK malting barley supply chain*. 12th Congress of the European Association of Agricultural Economists—EAAE 2008.

Li, D., Wang, X., Chan, H. K., & Manzini, R. (2014). Editorial—Sustainable food supply chain management. *International Journal of Production Economics, 152*, 1–8.

Mercer, D. G. (2011). Challenges facing development within the Agri-food sector of sub-Saharan Africa. 11th international congress on engineering and food (ICEF11). *Procedia Food Science, 1*, 1861–1866.

Mishra, D., & Satapathy, S. (2020a). MCDM approach for mitigation of flooding risks in Odisha (India) based on information retrieval. *International Journal of Cognitive Informatics and Natural Intelligence, 14*(2), 77–91. https://doi.org/10.4018/IJCINI.2020040105

Mishra, D., & Satapathy, S. (2020b). An integrated SWARA, QFD and ISM approach for agricultural injuries in India. *International Journal of Decision Support System Technology, 12*(2), 1–24. https://doi.org/10.4018/IJDSST.2020040101

Moomaw, W., Griffin, T., Kurczak, K., & Lomax, J. (2012). The critical role of global food consumption patterns in achieving sustainable food systems and food for all. In *A UNEP discussion paper, United Nations environment Programme*. Division of Technology, Industry and Economics.

Myers, R. J., Sexton, R. J., & Tomek, W. G. (2010). A century of research on agricultural markets. *American Journal of Agricultural Economics, 92*, 376–403.

Naik, G., & Suresh, D. N. (2018). Challenges of creating sustainable agri-retail supply chains. *IIMB Management Review, 30*, 270–282.

Narayanan S, Gulati A (2002) *Globalization and the smallholders: A review of issues, approaches, and implications*. MSSD Discussion Paper No. 50, International Food Policy Research Institute (IFPRI): Washington, DC, USA. Available online: http://ebrary.ifpri.org/cdm/ref/collection/p15738coll2/id/125154. Accessed on 8 Nov 2020.

Narrod, C., Roy, D., Okello, J., Avendano, B., Rich, K., & Thorat, A. (2009). Public-private partnerships and collective action in high value fruit and vegetable chains. *Food Policy, 34*, 8–15.

Olesen, J., & Bindi, M. (2002). Consequences of climate change for European agricultural productivity, land use and policy. *European Journal of Agronomy, 16*, 239–262.

Pang, Z., Chen, J., Zhang, Z., Chen, Q., Zheng, L. (2010). *Global fresh food tracking service enabled by wide area wireless sensor network*. https://doi.org/10.1109/SAS.2010.5439425.

Reardon, T. (2000). Agroindustrialization, globalization, and international development: An overview of issues, patterns, and determinants. *Agricultural Economics, 23*, 195–205.

Ricciardi, V., Ramankutty, N., Mehrabi, Z., Jarvis, L., & Chookolingo, B. (2018). How much of the world's food do smallholders produce? *Global Food Security, 17*, 64–72.

Ruben, R., Slingerland, M., & Nijhoff, H. (2006). Agro-food chains and networks for development—Issues, approaches and strategies. In R. Ruben, M. Slingerland, & H. Nijhoff (Eds.), *Agro-food chains and networks for development* (pp. 1–25). Springer.

Runge, C. F., Senauer, B., Pardey, P. G., & Rosegrant, M. W. (2003). *Ending hunger in our lifetime: Food security and globalization*. Washington, DC, USA.

Saccomandi, V., & van der Ploeg, J. (1998). *Agricultural market economics: A neo-institutional analysis of the exchange, circulation and distribution of agricultural products*. Assen, The Netherlands.

Searchinger, T., Hanson, C., Ranganathan, J., Lipinski, B., Waite, R., Winterbottom, R., Dinshaw, A., Heimlich, R. (2013) *Creating a sustainable food future: Interim findings*. A menu of

Solutions to Sustainably Feed More than 9 Billion People by 2050. World Resources Institute (WRI): Washington, DC, USA.

Shiferaw, B., Hellin, J., & Muricho, G. (2011). Improving market access and agricultural productivity growth in Africa: What role for producer organizations and collective action institutions? *Food Security, 3*(4), 475–489.

Shiferaw, B., Hellin, J., & Muricho, G. (2016). Markets access and agricultural productivity growth in developing countries: Challenges and opportunities for producer organizations. In J. Bijman, R. Muradian, & J. Schuurman (Eds.), *Economic Democratization, Inclusiveness and Social Capital*. Edward Elgar Publishing.

Shivakumar, D. (2016). *Agriculture: Time to add value to the agriculture value chain*. Business Today, 17th January 2016.

Sodano, V., Sassi, M., & Marchini, A. (2010). *Economia Agroalimentare: Mercati e Politiche*. Milan, Italy.

Stanujkic, D., Karabasevic, D., & Zavadskas, E. K. (2015). A framework for the selection of a packaging design based on the SWARA method. *Inzinerine Ekonomika-Engineering Economics, 26*(2), 181–187.

Sundar, I. (2016). Agribusiness scope, opportunities and challenges in India. *EPRA International Journal of Economic and Business Review, 4*(7), 171–178.

Swanson, B., & Rajalahti, R. (2010) *Strengthening agricultural extension and advisory systems: Procedures for assessing, transforming, and evaluating extension systems*.

Teklewold, H., Kassie, M., & Shiferaw, B. (2013). Adoption of multiple sustainable agricultural practices in rural Ethiopia. *Journal of Agricultural Economics, 64*(3), 597–623.

Trebbin, A. (2014). Linking small farmers to modern retail through producer organizations – Experiences with producer companies in India. *Food Policy, 45*, 35–44.

Tropp, D., Ragland, E., Barham, J. (2008). The dynamics of change in the U.S. food marketing environment. Agriculture handbook 728-3. U.S. Department of Agriculture Agricultural Marketing Service Marketing Services Program. July 2008.

UNCTAD. (2006). *Promoting TNC–SME linkages to enhance the productive capacity of developing countries" firms: A policy perspective*. Accessed on November 15, 2020 from http://unctad.org/en/docs/c3d75_en.pdf.

USDA. (2011). Available online at http://www.csrees.usda.gov/qlinks/extension.html. Accessed 14 Nov 2020.

Vermeulen, S. J., Campbell, B. M., & Ingram, J. S. I. (2012). Climate change and food systems. *Annual Review of Environment and Resources, 37*, 195–222.

Wollni, M., & Zeller, M. (2007). Do farmers benefit from participation in specialty markets and cooperatives? The case of coffee Marketing in Costa Rica. *Agricultural Economics, 37*, 243–248.

Wollni, M., Lee, A. R., & Thies, E. J. (2010). Conservation agriculture, organic marketing, and collective action in the Honduran hillsides. *Agricultural Economics, 4*, 373–384.

World Bank. (2007). *World development report 2008: Agriculture for development*. Washington, DC, USA.

World Bank. (2015). *Agriculture and poverty reduction*. Accessed on November 18, 2020 from https://www.worldbank.org/en/topic/agriculture

World Bank Group. (2016). *Global monitoring report 2015/2016: Development goals in an era of demographic change*. Washington, DC, USA.

Yakovleva, N. (2007). Editorial introduction: Measuring the sustainability of the food system. *Journal of Environmental Policy and Planning, 9*, 1–3.

Yazdani, M., Zolfani, S. H., & Zavadskas, E. K. (2016). New integration of MCDM methods and QFD in the selection of green suppliers. *Journal of Business Economics and Management, 17*(6), 1097–1113. https://doi.org/10.3846/16111699.2016.1165282

Zolfani, S. H., & Bahrami, M. (2014). Investment prioritizing in high tech industries based on SWARA-COPRAS approach. *Technological and Economic Development of Economy, 20*(3), 534–553. https://doi.org/10.3846/20294913.2014.881435

Zolfani, S. H., & Chatterjee, P. (2019). Comparative evaluation of sustainable design based on step-wise weight assessment ratio analysis (SWARA) and best worst method (BWM) methods: A perspective on household furnishing materials. *Symmetry, 11*(1), 74. https://doi.org/10.3390/sym11010074

Zolfani, S. H., Yazdani, M., & Zavadskas, E. K. (2018). An extended stepwise weight assessment ratio analysis (SWARA) method for improving criteria prioritization process. *Soft Computing*. https://doi.org/10.1007/s00500-018-3092-2

Chapter 7
Findings and Future Implications

7.1 Introduction

The "Internet of Things (IoT)" is a computational-supported information-system (Gubbi et al., 2013; Raiwani, 2013) that can be used for a variety of applications, which not only can help in reducing the physical workloads of farmers, but also can help in the ultimate reduction of farmers' mental stress levels in the workplaces. Other IT-based techniques are different from the IoT-based techniques with regard to their embedded and ubiquitous characteristics diffusing in daily-lives of every individual (Mohammed & Esmail, 2015). The IoT-based devices comprise of interactive elements, sensors and/or even complex-elements through well-defined communication-interfaces, which can share data apart from appropriate communications with the outer-world through specially designed network-gateways (Chen et al., 2011; Zachariah et al., 2015).

For the modernization of agricultural sectors, IoT has been offering a lot of innovative-solutions (Farooq et al., 2020), which include: monitoring-system for risks of landslide (Giorgetti et al., 2016); monitoring of soil, wind, water and atmosphere over larger areas (Zheng et al., 2016); monitoring of different environment-conditions for the creation in "dense and real time maps" for noise-levels, air, water-pollution, temperature, and damaging-radiations (Hachem et al., 2015; Torres-Ruiz et al., 2016); message alerts to authorities regarding collected data on different environment-parameters (Liu et al., 2013); use of wireless-sensors technology in a plant-nursery (Junaid, 2009); poly-house monitoring system making use of carbon-dioxide, temperature, humidity, and light-detection modules that were based on wireless-sensors (Song et al., 2012); monitoring of different agricultural-parameters by the use of "GPS technology and ZigBee protocol" (Satyanarayana & Mazaruddin, 2013); productivity enhancement based on real-time rice-crop monitoring-system (Sakthipriya, 2014); mitigation of the crop-loss risk and enhancement in crop-productivity by the use of crop monitoring-system for

S. Satapathy et al., *Innovation in Agriculture with IoT and AI*,
Innovation, Technology, and Knowledge Management,
https://doi.org/10.1007/978-3-030-88828-2_7

collecting information on temperature and rainfall (Rajesh, 2011); blue-tooth based system using micro-controller to monitor different agricultural-parameters (Shaobo et al., 2010); and monitoring of different environment-conditions, such as temperature, humidity, pressure and sunlight by the use of smart sensing-platform that was based on 'ZigBee' (Haefke et al., 2011); respectively.

Researches have been expanding around the globe in monitoring and fetching information with respect to crops' status in various stages by proposing different types of methods as well as architectures supported with a wide-variety of equipment. Different manufactures are engaged to provide a variety of sensors, robots, unmanned aerial-vehicles, communication-devices, and other heavy-machinery for effective execution of available data. In addition, different governmental-bodies, commissions, as well as food and agriculture organizations are in active development of guidelines and polices for observing and regulating the uses of newer and latest technologies for the maintenance of food and environment-safety (Thea et al., 2017). One of the major concern is drought that limits the crop-yield productivity and most of the regions in the world have been facing with such challenging issues. However, in order to deal with such issues, particularly in case of rural-areas, the use of remote-sensing devices is increasing to obtain the soil-moisture data frequently which helps in analyzing the agriculture-based drought. For these purposes, the use of "Soil-Moisture and Ocean-Salinity (SMOS) satellite" has been increasing since its launch in 2009, which helps in providing the global "soil-moisture maps" for every 'one to two days'. Moreover, Vågen et al. (2016) have used the "moderate resolution-imaging spectroradiometer (MODIS)" sensor to map different soil-functional properties for estimating the land-degradation risks for "sub-Saharan Africa". The "Normalized-Difference Vegetation-Index (NDVI)" can be used for the aerial and/or satellite images for monitoring the status of crops' nutrient (Benincasa et al., 2018; Liu et al., 2018). Different recent technologies that contribute to IoT-based smart-fertilization include: Geo-mapping (Suradhaniwar et al., 2018); Variable-Rate-Technology (Bruno et al., 2016; Colaço & Molin, 2017); and GPS-guided automated fertilizing-system (Khan et al., 2018); and crop-irrigation monitoring-systems (Briones et al., 2018; Raut et al., 2017; Villarrubia et al., 2017); respectively.

Agricultural intensification has the tendency to increase food availability to a greater-extent. However, adverse environment-related impacts occur due to this, such as increase in reactive-nitrogen over supply, organic processing of land as well as water, emissions of green-house gas and biodiversity-losses (Canfifield et al., 2010; Evenson & Gollin, 2003; IAASTD, 2009; Pingali, 2012). Induction of organic agriculture helps to improve the food-system's sustainability whose positive performance against a range of environment-related indicators has been reported widely (Schader et al., 2012; Reganold & Wachter, 2016). However, it results in low-yields (Seufert et al., 2012), and thus requires more land-areas for the production of comparable outputs as that of conventional production-systems. Moreover, the environment-related benefits of inducing organic agriculture are very few, if measured per unit-product than per unit-area (Tuomisto et al., 2012). Furthermore, the organic agriculture's ability to sustainably feed the world has been a challenge and the discussion of the authors have been limited to the consequence of low-yields by considering availability of nutrients (Connor, 2008; Connor, 2013).

In the recent developments in agriculture, the use of cloud computing has been playing a vital-role for storage of data. With the request of farmers to the cloud services, they can get the benefit of the cloud storage for storing the agricultural information as well as data, and the farmers also can get permission for accessing and analyzing the data. Since the use of "data mining" concept helps in getting useful information from a set of input-data, so further the data can be processed as well as analyzed by using big data "MapReduce" technique. Where, the map-function can be used for filtering as well as sorting performances with the reduce-function to initiate the summary-based operational performances. So, the MapReduce techniques can be useful in predictive analysis aspects. Rajeswari et al. (2018) have proposed a cloud-based 'big data analytics' concept for analyzing of agricultural field's data (Fig. 7.1). Initially with the use of pre-processing step, the noisy-values were eliminated with subsequent processing of data by using the discrimination-method. The use of categorization process was for the conversion of data into standard-format, and further the attributes' selection-method was used for important attributes' subset selection with subsequent application of classification-methods. Then with the use of MapReduce, the association-rule algorithm was applied for predicting the crop-patterns. This proposed model was recommended in providing an effective and smarter solution to the farmers for considerable improvement in crop-yields with lesser associated costs. Moreover, the present day technological advancements have leaded to a number of emerging and potential opportunities for monitoring of health aspects of the farming-community in terms of their body-parameters, which can help in providing their body state conditions with respect to some performing tasks or under a varying atmospheric conditions. With the help of "Raspberry pi" coupled to some body-parameter's sensing elements and displaying device (Fig. 7.2a) or by using the commercially available

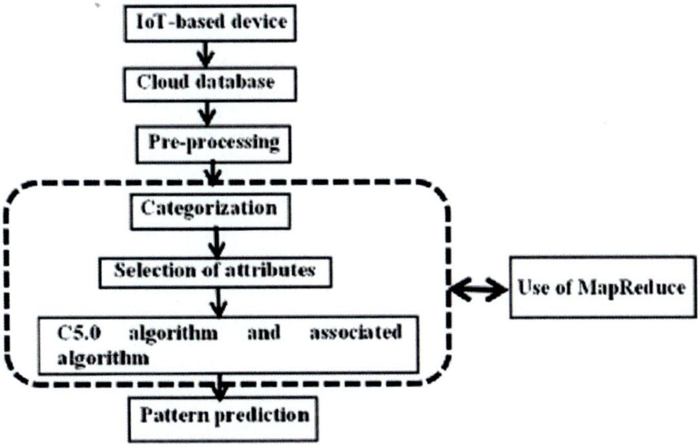

Fig. 7.1 Model for prediction of crop-patterns

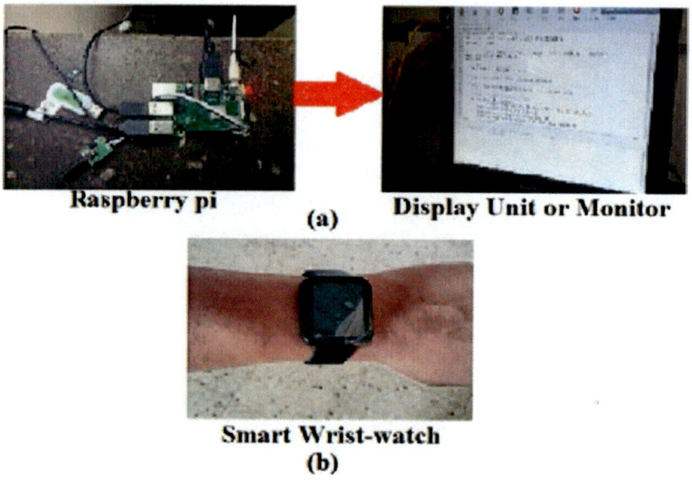

Fig. 7.2 Smart measuring devices for human body-parameters

wearable smart wrist-watches capable of indicating the body-parameters like "body-temperature and blood-pressure levels" (Fig. 7.2b), the existing physical-state of individuals can be predicted.

7.1.1 Strength, Weakness, Opportunities and Threats (SWOT) Analysis of Technologically Advanced Devices Over Existing Conventional Agricultural Systems and Techniques

In view of the "strength, weakness, opportunities and threats (SWOT)" of the agricultural devices, the SWOT analysis of both the technologically advanced devices as well as conventional agricultural systems and techniques were done separately with regard to their suitability and capability in workload reduction as well as measurement, mental stress reduction as well as measurement, and safety in workplaces as well as other safety-levels, respectively. It was found that the strengths associated with the conventional agricultural systems and techniques with regard to their suitability and capability in workload reduction as well as measurement included (Table 7.1): "Conventional tools are widely available; More and easier accessibility to farmers; Lesser costs; and Easy to operate". However, the opportunities included: "The tools and machines to be farmers' friendly; Easy to operate by all race groups irrespective of gender and sex; and Some tools can be manufactured locally".

Whereas, the strengths associated with the conventional agricultural systems and techniques with regard to their suitability and capability in mental stress reduction as well as measurement included: "Productivity in farms can be enhanced to some

Table 7.1 SWOT analysis of conventional agricultural systems and techniques with regard to workload reduction as well as measurement, mental stress reduction as well as measurement, and safety

Conventional agricultural systems and techniques		
Workload reduction as well as measurement	Strength	• Conventional tools are widely available • More and easier accessibility to farmers • Lesser costs • Easy to operate
	Weaknesses	• Higher associated costs of depreciation for most of the agricultural equipment • Difficulties to get some tools and machines in remote areas • Do not have any features of work-measurement
	Opportunities	• The tools and machines are farmers' friendly • Easy to operate by all race groups irrespective of gender and sex • Some tools can be manufactured locally
	Threats	• Drudgery level is more • Only some degree of workloads can be reduced
Mental stress reduction as well as measurement	Strength	• Productivity in farms can be enhanced to some extent • Lesser skilled-operators can operate the tools and small machines
	Weaknesses	• Do not have any features of mental stress measurement
	Opportunities	• Can be used in workload reduction to some degree which in turn deduces the associate mental stress • Play a vital role in income-generation to the farming-community
	Threats	• Increase in the development of mental stress among the farmers owing to the constrained benefits of utilization
Safety	Strength	• Easy to carry due to lesser weights of hand tools and other agricultural devices • No need of much training to operate the tools and machines • Use of protective devices in farming help the farmers to remain safe from adverse and dangerous health-impacts
	Weaknesses	• As work-related burdens are not reduced to larger extents, so possibilities of higher mental stress developments among the farmers • The manpower requirement is still more increasing stresses on farm owners • Do not completely eliminates the financial burdens by fulfilling the higher or expected "returns on investment" • Non-use of protective devices in farming cause the farmers to get exposed to adverse and dangerous health-impacts
	Opportunities	• Affordable by the small-scale farmers to purchase some of the smaller and most usable hand tools • Availing of existing subsidies on tools and machines that reduces the financial burdens among the farmers • Play a vital role in income-generation to the farming-community
	Threats	• Unsafe in most of the circumstances • Can cause in farm-related injuries • Can cause in musculoskeletal disorders of farmers

extent; and Lesser skilled-operators can operate the tools and small machines". However, the opportunities included: "Can be used in workload reduction to some degree which in turn deduces the associate mental stress; and Play a vital role in income-generation to the farming-community".

Similarly, the strengths associated with the conventional agricultural systems and techniques with regard to their safety included: "Easy to carry due to lesser weights of hand tools and other agricultural devices; No need of much training to operate the tools and machines; and Use of protective devices in farming help the farmers to remain safe from adverse and dangerous health-impacts". While, the opportunities included: "Affordable by the small-scale farmers to purchase some of the smaller and most usable hand tools; Availing of existing subsidies on tools and machines that reduces the financial burdens among the farmers; and Play a vital role in income-generation to the farming-community", respectively.

Further, the SWOT analysis of the technologically advanced devices revealed that the strengths associated with regard to their suitability and capability in workload reduction as well as measurement included (Table 7.2): "Some devices to be provided with the feature of working in wireless network; Wireless network based deices can be easily operated in remote areas; Lesser maintenance and depreciation costs; Ease of operations; Reduced workloads and drudgery-levels; and Remote areas can be covered easily". The associated opportunities included: "The tools and machines to be farmers' friendly; Easy to operate by all race groups irrespective of gender and sex; Wireless and remote operations can reduce the work-related burdens; Results in higher productivity; Results in higher incomes to the farming-communities with lesser efforts; Capable of covering more areas at lesser time; and additional features can be added for working loads (like pushing or pulling forces) measurement by the use of load-cells and sensors". However, the strengths associated with regard to mental stress reduction as well as measurement included: "Productivity in farms can be enhanced to greater extent; Lesser skilled-operators can operate the tools and small machines once they are provided with adequate training; and Reduction in work-related burdens reduce the associated mental stress among farmers". In addition to all these strengths, the inherent opportunities included: "Can be used in workload reduction to more extent which in turn deduces the associate mental stress; Play a vital role in income-generation to the farming-community; and Can be useful in the early prediction of agricultural problems including crop-yields, plant-diseases, soil and water quality, etc.

Similarly, the strengths associated with the technologically advanced devices with regard to their safety included: "Easy to carry due to lesser weights; More safe to operate the tools and machines; Provides greater protections to farmers from adverse and dangerous health-impacts; and More effective and efficient compared to the conventional agricultural systems". The possible and inherent opportunities included: "Results in higher productivity; Results in higher incomes to the farming-communities; Governmental agencies can provide adequate subsidies on advanced tools and machines that will reduce the financial burdens among the farmers; Provides more opportunities to preserve the farm-based products; and The farm-related injuries and accidents can be reduced to greater extent"; respectively.

Table 7.2 SWOT analysis of technologically advanced devices with regard to workload reduction as well as measurement, mental stress reduction as well as measurement, and safety

Technologically advanced devices		
Workload reduction as well as measurement	Strength	• Some devices are provided with the feature of working in wireless network • Wireless network based deices can be easily operated in remote areas • Lesser maintenance and depreciation costs • Ease of operations • Reduced workloads and drudgery-levels • Remote areas can be covered easily
	Weaknesses	• Lack of awareness among the farmers about the advanced tools and techniques • Lack of governmental supports in most of the developing countries
	Opportunities	• The tools and machines are farmers' friendly • Easy to operate by all race groups irrespective of gender and sex • Wireless and remote operations can reduce the work-related burdens • Results in higher productivity • Results in higher incomes to the farming-communities with lesser efforts • Capable of covering more areas at lesser time • Additional features can be added for working loads (like pushing or pulling forces) measurement by the use of load-cells and sensors
	Threats	• Need of additional training • Higher initial-costs compared to the conventional tools and machines
Mental stress reduction as well as measurement	Strength	• Productivity in farms can be enhanced to greater extent • Lesser skilled-operators can operate the tools and small machines once they are provided with adequate training • Reduction in work-related burdens reduce the associated mental stress among farmers
	Weaknesses	• Lack of awareness among the farmers about the advanced tools and techniques • Lack of governmental supports in most of the developing countries
	Opportunities	• Can be used in workload reduction to more extent which in turn deduces the associate mental stress • Play a vital role in income-generation to the farming-community • Can be useful in the early prediction of agricultural problems including crop-yields, plant-diseases, soil and water quality, etc.
	Threats	• Need of additional training • Higher initial-costs compared to the conventional tools and machines • Difficulties of small-scale farmers to access the available advanced tools and techniques

(continued)

Table 7.2 (continued)

Technologically advanced devices		
Safety	Strength	• Easy to carry due to lesser weights • More safe to operate the tools and machines • Provides greater protections to farmers from adverse and dangerous health-impacts • More effective and efficient compared to the conventional agricultural systems
	Weaknesses	• Lack of awareness among the farmers about the advanced tools and techniques • Lack of governmental supports in most of the developing countries
	Opportunities	• Results in higher productivity • Results in higher incomes to the farming-communities • Governmental agencies can provide adequate subsidies on advanced tools and machines that will reduce the financial burdens among the farmers • Provides more opportunities to preserve the farm-based products • The farm-related injuries and accidents can be reduced to greater extent
	Threats	• Need of additional training • Higher initial-costs compared to the conventional tools and machines • Difficulties of small-scale farmers to access the available advanced tools and techniques

7.2 Conclusion

The mental-health illness has been found to be associated with the physical workloads among the farming-community that mainly occurred due to their frustrations in agricultural works as well as more physical discomfort-levels. The development of mental-illness or mental stress can be considered to be taking place with the development of physical-discomforts. Technology adoption can be more relevant for better outcomes in agriculture with appropriate planning decisions for prevention of shortfalls, promotional activities and rehabilitation initiatives for the rural farming-communities.

As both male as well as female farmers are engaged in agricultural works with the probable possibilities of facing of a number of health-related problems including vomiting, nausea, headaches, skin-irritations, and so on, so adequate precautionary measures need to be under taken for smoother farm-related operations and activities. The farmers usually have to work in risky conditions like exposure to dirt, dust, sunlight, and chemicals while performing different farm-activities such as "weeding; harvesting; threshing; de-husking; cleaning; and crop-packaging"; respectively. Many of the farmers are not aware about the harmful-effect of the strong Sun's UV-light as well as other micro-organisms mixed with soils and of other dust-particles. Thus, the use of protective clothing can only protect the individual farmers getting exposed to such hazardous environment during their work-activities. Moreover, in order to safeguard and maintain the nutritional-levels of agricultural produces, adequate precautionary measures need to be given with the safe storage of such produces.

The farmers focus must be on better, smarter in addition to more efficient crop-growing methodologies for meeting the increasing food-demands of the accelerative world's population. Development of newer methods to improve crop-yields and adoption of innovative techniques in farming will help in considerable reduction of work-related burdens and thus, the associated disorders in mental-condition.

In view of the existing and ongoing problems in the global agriculture and among the farming-communities, this research work has considered all the farm-related aspects leading to physical and work related loads with subsequent mental ill-health among farmers. and highlighted the roles of different technologies for making the agriculture more efficient and smarter to meet the emerging expectations with lesser work burdens and associated work or mental stresses. Based on the discussions and findings, it can be concluded that adoption of technologically advanced techniques like IoT-based devices in the global farming sectors can help in maximizing production-levels and in the reduction or elimination of most of the existing problems in agriculture. The following recommendations were made on the basis of the outcomes of this research work.

A. For the prevention of uncertainties, problems and challenges in agriculture to protect farmers from the impacts on excessive physical-works and mental ill-health:

- Dividing the bigger and complex work-activities in smaller and shorter activities;
- Co-producing the national-vision for the future of agricultural-sectors;
- Agricultural progression through smarter working environment;
- Streamlining the regulatory as well as administrative processes in agriculture and other farming-activities;
- Ensuring of effective implementations of new approaches to support the digital-inclusion of farmers and their community;
- Creating awareness of workplace health-related programme and other initiatives among farmers;
- Facilitating of partnership-based working culture and integrating both governmental as well as non-governmental support-agencies for the well-being of the farming-community;
- The farmers should use hats and other clothes to protect them from heat and rain-water exposure, and the old age as well as children should avoid working in hot climactic-conditions;
- Creating awareness of the benefits of utilizing the protective and safety equipment in faring.

B. For the promotion of good mental-health and well-being among the farmers and their community:

- Supporting the farming-sectors to address the associated problems and hindrances to seek health or other agricultural advice and supports;
- Increasing the farmers' mental-health awareness in addition to the working-agencies with farmers;
- Initiating evidence-based approaches for mental-health promotions and resilience-building among the farmers;
- Governmental assistance to provide necessary political and other leadership supports during the agricultural-crisis periods;
- Governmental assistance to take necessary actions on price-regulation for agricultural produces.

7.3 Future Scope

The expectations of the farming-communities and their stress-levels have increased in measurable ways with the conventional ways of farming culture and practices. This has occurred owing to the work-related pressures and extrinsic stress-dimensions including nonproductive work practices and techniques, price-fluctuations, workload intensification, and agricultural regulation's changes. These have posed a serious threatening to the global farming-communities with the feelings of powerless to influence or control leading to their higher depression-rates, which has provided a major area for policy-intervention. The work-related aspects of farming-stress in turn provide the need for policy formulations and appropriate prevention strategies with regard to the negative consequences as well as impacts on farmers' health and well-being. The entire farming-communities can be greatly benefited by the use of ore technically advanced tools and equipment in the agriculture that in turn will reduce the work-burdens and other associated work-stresses leading to ill mental-health conditions. With adequate governmental supports and awareness creation among the farmers regarding the use of wireless network based control and monitoring of agricultural activities, considerable improvements in the global agri-sectors can be accomplished.

With the rapid developments and extensive promotions of computers and internet technologies, most of the users nowadays are more and more dependent on networks. The traditional information management systems in agricultural applications have been unable to meet the actual needs and large number of accesses with faster storage capabilities. In this regard, the IoT-based technology can be more suitably and effectively utilized in the farming-sectors.

The stress-levels in farming vary for both large and small scale farmers depending on their work culture, involved-activities and the way of work-executions with the existing and emerging complexities in farming. However, the small-scale farmers normally struggle with their intensified workloads, while large-scale farmers need to comply with the growing food demands for their financial stability. The role of supporting agencies plays a vital role to help these farming-communities by offering a wider range of proactive as well as responsive services. The locally-based supports can be used for the needy rural farmers by establishing ore transparent and effective local-networks, however through partnership-based approaches, the access to a wider range of services through some single-point contact can be more successful in most of the faring-sectors. Critically, the supports need to be multi-dimensional in nature that can reflect the farm-related stressors with their adverse impacts among the farming-communities. Moreover, the global farming-communities can be greatly benefited by providing agriculture focused stress as well as safety management training, allowing for more active participation of farmers, guidance role's development in order to access the health as well as safety information for supporting the farmers, advisory services' building to help the farmers for the development of alternative and flexible working-structures, and by expanding more awareness in addition to accessibility of both the existing and newer initiatives with regard to policies and technological advancements. Furthermore, the respective governmental bodies can take different endeavors to streamline the agencies associated with farming regulations; to simplify, coordinate and improve the administrative system's quality in addition to their supports; to provide robust information on impending policy-changes; and to provide adequate subsidies to the farmers for availing the technical and innovative benefits in terms of tools and equipment for less stressful and more productive agriculture.

References

Benincasa, P., Antognelli, S., Brunetti, L., Fabbri, C., Natale, A., Sartoretti, V., & Vizzari, M. (2018). Reliability of NDVI derived by high resolution satellite and UAV compared to in-field methods for the evaluation of early crop N status and grain yield in wheat. *Experimental Agriculture, 54*(4), 604–622.

Briones, A. G., Castellanos-Garzón, J. A., Martín, Y. M., Prieto, J., & Corchado, J. M. (2018). A framework for knowledge discovery from wireless sensor networks in rural environments: A crop irrigation systems case study. *Wireless Communications and Mobile Computing, 2018*, 14.

Bruno, B., Benjamin, D., Davide, C., Andrea, P., Francesco, M., & Luigi, S. (2016). Environmental and economic benefits of variable rate nitrogen fertilization in a nitrate vulnerable zone. *Science of the Total Environment, 545-546*, 227–235.

Canfifield, D., Glazer, A., & Falkowski, P. (2010). The evolution and future of earth's nitrogen cycle. *Science, 330*, 192–196.

Chen, H., Jia, X., & Li, H. (2011). A brief introduction to IoT gateway. In *IET international conference on communication technology and application (ICCTA 2011)* (pp. 610–613). IET. https://doi.org/10.1049/cp.2011.0740. http://ieeexplore.ieee.org/stamp/stamp.jsp?tp=&arnumber=6192937&isnumber=6192814

Colaço, A. F., & Molin, J. P. (2017). Variable rate fertilization in citrus: A long term study. *Precision Agriculture, 18*(2).

Connor, D. J. (2008). Organic agriculture cannot feed the world. *Field Crops Research, 106*, 187–190.

Connor, D. (2013). Organically grown crops do not a cropping system make and nor can organic agriculture nearly feed the world. *Field Crops Research, 144*, 145–147.

Evenson, R. E., & Gollin, D. (2003). Assessing the impact of the green revolution, 1960–2000. *Science, 300*, 758–762.

Farooq, M. S., Riaz, S., Abid, A., Tariq Umer, T., & Zikria, Y. B. (2020). Role of IoT Technology in Agriculture: A systematic literature review. *Electronics, 9*, 319. https://doi.org/10.3390/electronics9020319

Giorgetti, A., Lucchi, M., Tavelli, E., Barla, M., Gigli, G., Casagli, N., & Dardari, D. (2016). A robust wireless sensor network for landslide risk analysis: System design, deployment, and field testing. *IEEE Sensors Journal, 16*, 6374–6386.

Gubbi, J., Buyya, R., Marusic, S., & Palaniswami, M. (2013). Internet of things (iot): A vision, architectural elements, and future directions. *Future Generation Computer Systems, 29*(7), 1645–1660.

Hachem, S., Mallet, V., Ventura, R., Pathak, A., Issarny, V., Raverdy, P. G., Bhatia, R. (2015) Monitoring noise pollution using the urban civics middleware. In *Proceedings of the 2015 IEEE first international conference on big data computing service and applications, Redwood City, CA, USA, 30 March–2 April 2015*; pp. 52–61.

Haefke, M., Mukhopadhyay, S. C., Ewald, H. (2011). A Zigbee based smart sensing platform for monitoring environmental parameters. In *Proceedings of the 2011 IEEE international instrumentation and measurement technology conference, Binjiang, China, 10–12 May 2011*; pp. 1–8.

IAASTD. (2009). *Agriculture at a crossroads: The global report*. Island Press.

Junaid, A. (2009). *Application of modern high performance networks* (pp. 120–129). Bentham Science Publishers Ltd.

Khan, N., Medlock, G., Graves, S., & Anwar, S. (2018). *GPS guided autonomous navigation of a small agricultural robot with automated fertilizing system*. SAE Technical Paper.

Liu, Z., Huang, J., Wang, Q., Wang, Y., Fu, J. (2013) Real-time barrier lakes monitoring and warning system based on wireless sensor network. In *Proceedings of the 2013 fourth international conference on intelligent control and information processing (ICICIP), Beijing, China, 9–11 June 2013*; pp. 551–554.

Liu, H., Wang, X., & Bing-kun, J. (2018). Study on Ndvi optimization of corn variable fertilizer applicator. *Agricultural Engineering, 56*(3), 193–202.

Mohammed, F. H., & Esmail, R. (2015). Survey on iot services: Classifications and applications. *International Journal of Science and Research, 4*, 2124–2127.

Pingali, P. L. (2012). Green revolution: Impacts, limits, and the path ahead. *Proceedings of the National Academy of Sciences of the United States of America, 109*, 12302–12308.

Raiwani, Y. (2013). Internet of things: A new paradigm. *International Journal of Scientific and Research Publications, 3*(4), 323–326.

Rajesh, D. (2011). Application of spatial data mining for agriculture. *International Journal of Computers and Applications, 15*, 7–9.

Rajeswari, S., Suthendran, K., & Rajakumar, K. (2018). A smart agricultural model by integrating IoT, Mobile and cloud-based big data analytics. *International Journal of Pure and Applied Mathematics, 8*, 365–370.

Raut, R., Varma, H., Mulla, C., & Pawar, V. R. (2017). Soil monitoring, fertigation, and irrigation system using IoT for agricultural application. *Intelligent Communication and Computational Technologies*, 67–73.

Reganold, J. P., & Wachter, J. M. (2016). Organic agriculture in the twenty-first century. *Nature Plants, 2*, 1–8.

Sakthipriya, N. (2014). An effective method for crop monitoring using wireless sensor network. *Middle-East Journal of Scientific Research, 20*, 1127–1132.

Satyanarayana, G. V., Mazaruddin, S. D. (2013). Wireless sensor based remote monitoring system for agriculture using ZigBee and GPS. In *Proceedings of the conference on advances in communication and control systems-2013, Makka Wala, India, 6–8 April 2013*.

Schader, C., Stolze, M., & Gattinger, A. (2012). In J. I. Boye & Y. Arcand (Eds.), *Green Technologies in Food Production and Processing* (pp. 183–210). Springer.

Seufert, V., Ramankutty, N., & Foley, J. A. (2012). Comparing the yields of organic and conventional agriculture. *Nature, 485*, 229–232.

Shaobo, Y., Zhenjianng, C., Xuesong, S., Qingjia, M., Jiejing, L., Tingjiao, L., Kezheng, W. (2010) The application of bluetooth module on the agriculture expert System. In *Proceedings of the 2010 2nd international conference on industrial and information systems, Dalian, China, 10–11 July 2010*; Volume 1, pp. 109–112.

Song, Y., Ma, J., Zhang, X., & Feng, Y. (2012). Design of wireless sensor network-based greenhouse environment monitoring and automatic control system. *Journal of Network, 7*, 838.

Suradhaniwar, S., Kar, S., Nandan, R., Raj, R., & Jagarlapudi, A. (2018). Geo ICDTs: Principles and applications in agriculture. In G. Reddy & S. Singh (Eds.), *Geospatial Technologies in Land Resources Mapping, monitoring and management*. Geotechnologies and the Environment, Springer 21.

Thea, K., Martin, C., Jeffrey, M., Gerhard, E., Dimitrios, Z., Edward, M., & Jeremy, P. (2017). Food safety for food security: Relationship between global mega trends and developments in food safety. *Trends in Food Science & Technology, 68*, 160–175.

Torres-Ruiz, M., Juárez-Hipólito, J. H., Lytras, M. D., Moreno-Ibarra, M. (2016). Environmental noise sensing approach based on volunteered geographic information and spatio-temporal analysis with machine learning. In *Proceedings of the international conference on computational science and its applications, Beijing, China, 4–7 July 2016*; pp. 95–110.

Tuomisto, H., Hodge, I., Riordan, P., & Macdonald, D. (2012). Does organic farming reduce environmental impacts? –a meta-analysis of European research. *Journal of Environmental Management, 112*, 309–320.

Vågen, T.-G., Winowiecki, L. A., Tondoh, J. E., Desta, L. T., & Gumbricht, T. (2016). Mapping of soil properties and land degradation risk in Africa using MODIS reflectance. *Geoderma, 263*, 216–225.

Villarrubia, G., De Paz, J. F., De La Iglesia, D. H., & Bajo, J. (2017). Combining multi-agent systems and wireless sensor networks for monitoring crop irrigation. *Sensors, 17*(8), 1775.

Zachariah, T., Klugman, N., Campbell, B., Adkins, J., Jackson, N., & Dutta, P. (2015). The internet of things has a gateway problem. In *Proceedings of the 16th international workshop on Mobile computing systems and applications, HotMobile '15* (pp. 27–32). ACM. https://doi.org/10.1145/2699343.2699344

Zheng, R., Zhang, T., Liu, Z., & Wang, H. (2016). An EIoT system designed for ecological and environmental management of the Xianghe Segment of China's Grand Canal. *International Journal of Sustainable Development & World Ecology, 23*, 372–380.

Index

S. Satapathy et al., *Innovation in Agriculture with IoT and AI*, Innovation, Technology, and Knowledge Management,
https://doi.org/10.1007/978-3-030-88828-2